Applying a Systems Thinking the Construction Industry

This book aims to shed light on why it is that so many well-meaning initiatives and government white papers have failed to have the expected impact in transforming the UK construction industry. Using the UK housing sector as a case study, Mike Siebert applies a Systems Thinking approach to tackling some of the shared 'Wicked Problems' faced by an industry that urgently needs to boost its productivity levels, build more sustainably and affordably, and generally improve its working practices.

In an accessible and easy to read style, Siebert challenges the overall decision making and problem-solving approach adopted by the industry and seeks to put Systems Thinking front and centre to consider the core issues from multiple perspectives. Initially outlining the key stakeholders and the drivers and barriers to change, he then introduces Systems Thinking and explains using numerous examples of known issues what this approach could achieve.

His central aim is to show how, if a Systems Thinking approach were to be applied to the UK housing industry's problems, many of them could be resolved to the benefit of all the parties involved – government, housebuilders, material suppliers, the warranty industry, the design industry and the end users. These are shared problems, and they require shared solutions, but without first understanding these complex problems from the perspectives of all parties that need to benefit from the solutions being proposed, it is unlikely that those solutions will achieve the level of engagement needed for them to successfully meet their objectives.

Mike Siebert is a practising architect, active researcher and founder of Ecologic Homes Consultancy Ltd. He has advised Nottingham City Council, lectured and tutored at the University of Nottingham and Nottingham Trent University, and worked with the Construction Innovation Hub to help deliver the UK government's latest initiatives aimed at transforming the construction industry, including the Platform Design Approach and the Value Toolkit.

Applying a Systems Thinking Approach to the Construction Industry

Mike Siebert

Routledge
Taylor & Francis Group

LONDON AND NEW YORK

Designed cover image: Cover photograph by Mike Siebert of Chichu Art Museum by Tadao Ando

First published 2024
by Routledge
4 Park Square, Milton Park, Abingdon, Oxon OX14 4RN

and by Routledge
605 Third Avenue, New York, NY 10158

Routledge is an imprint of the Taylor & Francis Group, an informa business

British Library Cataloguing-in-Publication Data
A catalogue record for this book is available from the British Library

Library of Congress Cataloging-in-Publication Data
Names: Siebert, Mike, author.
Title: Applying a systems thinking approach to the construction industry / Mike Siebert.
Description: Abingdon, Oxon ; New York, NY : Routledge, 2023. |
Includes bibliographical references and index.
Identifiers: LCCN 2023018835 (print) | LCCN 2023018836 (ebook) |
ISBN 9781032366180 (hardback) | ISBN 9781003332930 (ebook)
Subjects: LCSH: Construction industry–Great Britain. |
Housing–Great Britain.
Classification: LCC HD9715.G72 S54 2023 (print) |
LCC HD9715.G72 (ebook) | DDC 338.4/76240941–dc23/eng/20230519
LC record available at https://lccn.loc.gov/2023018835
LC ebook record available at https://lccn.loc.gov/2023018836

ISBN: 978-1-032-36618-0 (hbk)
ISBN: 978-1-032-36059-1 (pbk)
ISBN: 978-1-003-33293-0 (ebk)

DOI: 10.1201/9781003332930

Typeset in Times New Roman
by Newgen Publishing UK

Contents

List of Figures   *vii*
Preface   *ix*
The Structure of the Book   *xii*

Introduction   1

PART I
What Are We Trying to Do? What's Stopping Us? How Could
We Approach It Differently?   7

1   Defining the Sectors   9

2   Drivers for Change: The Government Perspective on What
     the Industry Needs to Do   30

3   Barriers to Change: The Industry Perspective on What the
     Government Needs to Understand   48

4   The Methodology for Change   76

PART II
What 'Approaching It Differently' Could Achieve   109

5   Political Intervention: Changing the Environment That the
     'Industry Organism' Functions within   115

6   Industry Intervention: Changing How the 'Industry
     Organism' Functions within Its Environment   128

7   The Missing Tools: The Tools We Need, How to Build
    Them, and How to Promote Them   146

8   Applying the Tools to the Housing Industry   187

    Conclusion   202

    Index   *204*

Figures

1.1	Energy use across offices with EPC ratings from G to A	5
1.2	Industry sectors – top-level divisions	10
1.3	Industry drivers	11
1.4	Government agendas	13
1.5	Government motivators	15
1.6	Divisions within the housing industry	16
1.7	Supply chain divisions	20
1.8	Drivers within the design sector	23
1.9	Drivers amongst end users	25
1.10	Divisions within residential and non-residential	26
1.11	Breakdown of drivers across all sectors	27
2.1	Construction industry government white papers up to the Farmer Review	32
2.2	The adoption of OSM. Adapted from *Crossing the Chasm*	40
2.3	Construction 2025 targets	44
2.4	The hierarchy of barriers to adoption	45
3.1	How we access information	51
3.2	The cyclical adoption of off-site manufacturing	55
3.3	Barriers to the ability to adopt	56
3.4	The Farmer Review's ten symptoms	59
3.5	The options ahead	60
3.6	Root causes of inaction vs. Farmer's ten symptoms of poor performance	61
3.7	The construction industry's productivity deficit	64
4.1	The different flavours of Systems Thinking	79
4.2	Systems Thinking categorisations	80
4.3	Critical, Tame and Wicked Problems	84
4.4	Decision-making tree for innovators	89
4.5	The next level of questioning to establish the benefits	96
4.6	Example of benefits for the supplier	97
4.7	The next level of questioning to establish the barriers	99
4.8	The five sectors to be interrogated	100
4.9	Table of pros and cons for CLT	101

4.10   Decision-making tree for developers   104
4.11   The innovator's options   107
5.1   Scimagojr bubble chart of research areas   116
5.2   Scimagojr bubble chart breakdown for engineering   116
5.3   The box-ticking conundrum   118
6.1   The value of early info as delivered by BIM   132
6.2   Supply chain database model for mutual engagement   136
6.3   Alternative metric systems (not) in use   139
7.1   The Value Toolkit wheel   148
7.2   The Metriculator – Step 1, company profile   150
7.3   The Metriculator – Step 2, project costs   151
7.4   The Digital Cousin approach   154
7.5   The Metriculator – Step 3, 'behind the scenes' metric influences   156
7.6   The Metriculator – Step 4, metric weighting   157
7.7   The Metriculator – relationship model   158
7.8   An attempt to map the industry's various bodies   160
7.9   The three steps of Capability, Credibility and Compatibility   162
7.10   One central resource with multiple entry points   163
7.11   Stages in the decision-making process   167
7.12   Stage 1 – identify the known issues   168
7.13   Stage 2 – causes and consequences   169
7.14   Stage 3 – hierarchy of causes   170
7.15   Stage 4 – the addition of further causes   171
7.16   Stage 5 – hierarchical connections   172
7.17   The decision-making tool – known issues   179
7.18   The decision-making tool – causes and consequences   181
7.19   The decision-making tool framework   183
8.1   The housing crisis – the known issues   188
8.2   The housing crisis – causes and consequences   189
8.3   The options – confronting the barriers to bring about change.   192
8.4   Residual land value calculation   197

Preface

The last 'big read' relating to the UK construction industry was Mark Farmer's paper, 'Modernise or Die', in 2016. It was a hard-hitting, comprehensive analysis of the many issues the construction industry needed to confront, analogised as the symptoms of a dying patient. But as with the many other papers preceding it proposing radical changes, it has so far failed to have the impact it was expected and needed to achieve.

This book is not another attempt to prescribe what the construction industry needs to do to boost its productivity levels, build more sustainably and affordably, or generally improve its working practices. It is more an attempt to understand why it is that so many of these initiatives fail to result in any substantive change in mainstream practice. Whilst there are many theories for what lies behind this constraint on progress, often hinting at complacency within the industry as an underlying factor, it is an area of debate that continues to receive too little attention. Instead, new initiatives continue to arrive, prefaced with an acknowledgment of past failures and a veiled threat of what will befall the industry if, yet again, it continues to ignore or 'cherry pick' from the advice being given (Climate Assembly UK, 2020; Eclipse Research Consultants, 2009).

The initial proposition is a simple one. It is not that all these initiatives are necessarily wrong in their objectives, but that they too often fail to address how they are to be implemented on the ground once unveiled. If this were the total extent of the problem to be addressed, however, this would be a 'small read', but the problem we're facing is possibly not just this missing final step. A far more fundamental missing component in these 'change initiatives' is how they are being conceived from the outset, with the lack of attention given to their implementation being a consequence of that limited range of perspectives considered at those earlier stages. Any initiative primarily driven by government objectives – be that boosting productivity levels, reducing carbon emissions, or levelling up – needs those broader economic benefits to be translated into a language that those businesses being asked to change their working practices can relate to in terms of their own profitability. Sometimes this really is just an exercise in translation – a missing final step – but more often it can be shown to be that failure early on to consider a proposal's impact at an individual business level that renders an entire initiative unsellable and therefore unlikely to bring about any meaningful change.

The solution being proposed is also potentially simple, but not something that the construction industry is well equipped to implement, for reasons that will first need to be overcome. The industry's deeply embedded adversarial and fragmented nature is well documented (Bourn, 2001; Farmer, 2016; Integrated Project Initiatives Ltd., 2014) and often cited as a root cause for its many ills, but it is also where the change must begin if any solution is to stand a chance of adoption. Within that fragmented structure lie many disparate and often conflicting perspectives on what change means and how any particular change may or may not be of benefit to an individual business. It is these perspectives that need to be better understood and listened to alongside the dominant perspective of an initiative's instigator. What this might reveal is that the construction industry is not resistant to change as such, but to top-down change programmes, and it is the shortcomings of this model that must be addressed before wastefully fomenting any further initiatives.

The suggested vehicle for this is Systems Thinking, a methodology that originated as Systems Engineering within the space industry, where it was imperative that all possible outcomes were considered from all perspectives and then catered for prior to any launch. As an approach to complex problem solving, Systems Thinking has now been honed over many decades and across many industry sectors, most notably within the manufacturing industry. The 'case study' used here for assessing the potential of this approach for the construction industry is the sector most in need of that broader perspective, our housing industry. What makes housing a special case is the added level of direct societal impact it has on all our lives, and the consequential importance of that additional end user perspective, so often missing from the industry's decision-making process. A Systems Thinking approach aims to capture all perspectives and build a far more balanced understanding of where the benefits lie, what the motivations for change actually are, and ultimately ask the question, 'what's in it for me?' Successful interventions are dependent on there being an answer to that question from all those who are party to its implementation. And to be clear, this is not an attempt to give all parties an equal say and pretend that the industry could be run as a co-operative, but to recognise the importance and financial benefit to be gained, both individually and collectively, from understanding all those other perspectives and why, when not considered, their unmet concerns so often result in disappointingly low adoption rates.

Part of the challenge in writing a book that addresses those many sectors' differing perspectives is that it needs to be readily accessible on many levels without assuming or explaining too much. The format adopted here is therefore as important as the content in getting that message across, as writing such a book for only one audience – be that our own industry, government legislators, the end users or academia – would fail to expose the other sectors to the alternative viewpoints that they are not currently hearing. The objective here is to show the importance of understanding the industry's complex problems from those multiple perspectives before attempting to change the way in which it functions, and it aims to do this by comparing current and recent initiatives against theoretical alternatives that arise from taking a Systems Thinking approach. It is therefore important to point out that the proposed alternative solutions are both untested and secondary to the book's main objective, which is to change the way the industry approaches problem solving, as a precursor to – and catalyst for – changing the way in which it operates. In that respect, it can be seen as a plea for innovative decision making to be driven by an approach that focuses on everyone asking the right questions, as opposed to a select few dictating the 'right' answers, a concept that has proven to be central to all the propositions that are now put forward here for discussion.

The Structure of the Book

This book is in two parts – the first explains the proposed approach to decision making and why it is needed, based on evidence of how decisions are currently being made in the construction industry compared to how they are made in other industry sectors. The second part contrasts the low engagement levels achieved by recent initiatives with the potentially far better outcomes that could be achieved through some example initiatives based on implementing this different approach. The main objective is to encourage better decision making rather than proposing specific solutions, and to prove that there's a benefit to all parties in taking this approach by way of some key examples that all sectors can relate to. As in collaborative working itself, this endeavour will require there to be a direct benefit to everyone in reading about / engaging with what that benefit might look like for others, which has resulted in an approach that tries to view each scenario from multiple perspectives in turn.

The overall structure therefore requires each chapter to be read in the right order and ideally, in its entirety:

Introduction

The purpose of this book. To suggest why it is that the construction industry struggles to adopt change and how that could be remedied by working on its collective approach to complex problem solving.

Part I What Are We Trying to Do? What's Stopping Us? How Could We Approach It Differently?

A synopsis of the problems being faced, why we need to deal with them, the approaches we've tried so far, and how those initiatives could be presented or constructed differently to achieve better results.

Chapter 1 Defining the Sectors

A broad analysis of how the industry is structured, to explain why it cannot be treated as a unified entity but must be understood at a more granular level.

Chapter 2 Drivers for Change: The Government Perspective on What the Industry Needs to Do

An explanation of what is driving interventions from a government perspective and how that differs from the industry's many and varied drivers that dictate the decisions individual businesses make.

Chapter 3 Barriers to Change: The Industry Perspective on What the Government Needs to Understand

An explanation of the many reasons why change is difficult to both promote and adopt within an adversarial and fragmented industry.

Chapter 4 The Methodology for Change

How the decision to change must come from a knowledge not only of the benefits to be gained by all parties but also of the barriers to be overcome to make that transition possible.

Part II What 'Approaching It Differently' Could Achieve

A translation of the proposed Systems Thinking methodology into strategies for intervention together with some example solutions that can be derived from this approach.

Chapter 5 Political Intervention: Changing the Environment That the 'Industry Organism' Functions within

How government operates and how those ways of operating can be influenced by understanding the government's true motivations but also by explaining to government the industry's very different motivations.

Chapter 6 Industry Intervention: Changing How The 'Industry Organism' Functions within Its Environment – Changing the Message to Fit the Audience

How the many sectors within the construction industry need to be engaged at all levels before any change will be successfully adopted.

Chapter 7 The Missing Tools: The Tools We Need, How to Build Them, and How to Promote Them

What are the missing tools that the industry needs in order to reconnect with its own composite parts and work more collaboratively towards innovation and boosting its productivity levels?

Chapter 8 Applying the Tools to the Housing Industry

How these approaches could all be brought into play to fix the way in which our housing industry operates for the good of the economy, the environment and our society.

Conclusion

Why a Systems Thinking approach is necessary and why its methodologies have to be applied before any other measures can hope to succeed.

Introduction

It would appear the construction industry is in need of reform, and has been for some time. Eighty years is a long time to be making the same claims to no effect, but that is the reality of what has been happening, since the book *Building to the Skies: The Romance of the Skyscraper* was published in 1934. The industrialist Alfred Bossom, on his return from America, saw in Britain 'an adversarial and wasteful industry in which construction took too long, was too expensive and was not satisfactory for its clients'. He observed that the construction industry was just an industry like any other, and that if it were to adopt more organised, collaborative working arrangements this would lead to increased productivity, profits, investment and wages (Bossom, 1934).

Mark Farmer's seminal report, published in 2016 and entitled 'Modernise or Die', covered similar topics, but used some stronger language. The industry's ails are now likened to the symptoms of a dying patient, each of which are discussed in detail together with the familiar bullet-pointed recommendations outlining what needs to be done. But, as with all dying patients, there's now a 'deadline' and an urgency that means, this time, the findings of this report cannot be ignored.

What is missing from this comprehensive deep dive into the causes and consequences of the industry's many problems seems to be one fundamental question. Why is it that decades of government-commissioned reports, many referenced by Mark Farmer and held up as seminal pieces of research in their own right, were received by so much of the industry with such indifference? If the belief is that the threat of imminent death will be enough to make the industry – and government – sit up in bed and take these latest recommendations more seriously, that might be to make a dangerous presumption about where the root cause of this illness lies.

The purpose of this book is to consider these problems that absolutely do exist within the industry, but to suggest one more – the industry's inadequate approach to problem solving. The one thing these many government-initiated white papers have in common, beyond their failure to live up to their intended and expected transitional impact, is possibly their failure to fully understand the complexity of the industry they are all trying to influence. A detailed study

DOI: 10.1201/9781003332930-1

of previous government initiatives would suggest that it is often the case that the analysis of the problems being investigated is thorough and sound, but the proposals for how the prescribed remedies should be administered to actual businesses have not received the same level of attention. Sometimes this failure to understand the industry at a more granular level has led to some more fundamentally flawed solutions being proposed; on other occasions all that is now needed is a translation of those solutions into a language that can be understood by those who need to hear the message. Just answering one last question – 'what's in it for me?' – will either bridge that gap between disinterest and willing engagement, or expose that gap as being an unbridgeable void, and requiring a more fundamental rethink.

The first section in this book looks at that decision-making process as it currently exists through the lens of some recent initiatives, and attempts to show how the conclusions reached are predominantly focused on the benefit to be gained from the perspective of the originator, and how this might fail to engage those within the industry who need to enact the changes being proposed. It then goes on to define the industry in more detail in order to expose its complexity, the variety of drivers that need to be satisfied, and the potential for these to be in conflict, resulting in barriers to adoption. From this stems the proposition for a multi-perspective, Systems Thinking approach that recognises those motivational needs and builds a case for satisfying all of them as a prerequisite for the broad engagement that is essential for any innovation to succeed.

Part II then applies this approach to some of the known issues confronting the industry and some of the recent initiatives that have fallen short of their own expectations. By applying a Systems Thinking approach which re-runs the same analytical processes, but this time informed by a far broader range of perspectives representing all those involved, a different set of outcomes emerge. The purpose here is not primarily to promote these alternative solutions, but to promote a more informed approach to decision making by showing how the alternative solutions reached using this approach are beneficial to all parties and therefore far more likely to be adopted. Debating the validity of these alternative solutions, however, using the same Systems Thinking approach, would be a logical next step. New initiatives for transforming the construction industry continue to be written, but we really ought to be focusing first on this missing piece of the jigsaw before moving on to the next big picture.

Part I What Are We Trying to Fix – and What's Stopping Us?

Taking a step back is always a good way to start, but never an easy proposition to sell. Short termism is a criticism made of many governments, industries and businesses, and taking a long-term, broad look at a problem from behind the starting line is as unlikely to be well received, as is considering a problem from multiple perspectives. But as the urgency of the problems confronting the construction industry increases to near existential levels (Farmer, 2016, p. 5), the ability or willingness to do so diminishes even more, resulting in an even greater tendency to propose simple 'quick fixes' at increasingly grand scales. Resisting this is therefore a challenge in itself, but a necessary one if we are to make the right decisions and make any meaningful progress towards our intended goal.

So what is our goal? What is it we are trying to fix by transforming the way our construction industry operates? Again, Mark Farmer's comprehensive review catalogues the industry's many problems that need addressing, but the first question that comes into focus by taking that step back is 'who is this *we*'? This is important, because what 'we' are trying to fix depends entirely on who is being asked the question, which would suggest that the answer to the next question, 'what's stopping us?', is perhaps a failure to establish what all those different 'we perspectives' are on the purpose behind transforming the construction industry in the first place. The proposition being made here is that if there were a better understanding of those alternative perspectives, our initiatives might better reflect what businesses within the industry need to get out of transforming it, and as a result be more supportive of the initiatives being promoted. The first challenge is therefore one of defining the industry in its broadest sense to capture all the different perspectives that exist, including the government's, but only as one voice amongst many that need to be heard.

Who Are the 'We's? (Chapter 1)

The government; government departments; local authorities; the industry; the businesses within that industry; the employees within those businesses.

That is already six distinct interest groups, within which there are many more subdivisions to be found, and outside of which there are still more sectors that need to be included. Just as there are internal tensions between different government departments, there are also tensions within the construction industry, with its desire to build more affordably and sustainably, be more productive and limit its exposure to risk – all generating their own conflicting drivers. But as well as this, at an individual business level and across the whole supply chain, there are many more drivers that exist at a more prosaic level, right down to interpersonal decisions based on friendships and embedded knowledge that will often dictate how these broader decisions are taken.

What Are 'We' Trying to Fix? (Chapter 2)

This has to begin with understanding the dominant government perspective, before introducing the many other alternative points of view so often overlooked. And even this can take some time to decipher, as the higher up the hierarchy of decision making we travel, the less likely it is we will be privy to the genuine reasoning behind the decisions being taken.

The three main thrusts of government policy under the 2021 post-Covid 'Build Back Better' agenda were 'Net Zero', 'Global Britain' and 'Levelling Up'. Within these slogans can be found a desire to boost productivity through encouraging innovation and to spread this around the country – which could equally well be translated as a government needing to increase our GDP to strengthen our economy and secure its own re-election. All of these are legitimate, if sometimes subjective, ambitions, but they are still only the aims of one participant and defined in terms of benefit at a national political and economic level. What this process will add to the equation are the voices of those other stakeholders who need to be part of that conversation, and whose drivers are noticeably different.

What's Stopping Us? (Chapter 3)

Clearly we all benefit from a stronger economy as we also would from a fairer, more sustainable economy, but that is not necessarily how small businesses measure their success or make their decisions. Consequently many of the messages contained within these initiatives get 'lost in translation', or to be more accurate, never get translated into a language that businesses can relate to. To take a step back once more, before that translation can happen, there first needs to be a better understanding of the mindsets within the industry that government needs to be translating for. In many respects this is the final missing piece of the jigsaw that will make sense of the picture the government is trying to present, and it begs the question why it is that so little time and effort goes into the implementation of initiatives that have been so thoroughly researched and developed.

What Needs to Change? (Chapter 4)

It would be naïve to think that one missing piece needs dropping into place to complete the picture and uncork a tsunami of change across all these areas of concern – productivity, adoption of innovation, capturing of industry metrics, training and retention of skilled workers, etc. The fact that these missing perspectives remain missing would suggest that there is either a lack of appreciation of their relevance as part of this implementation stage or that they are not straightforward to capture. What clearly does not exist, however, is any evidence of there being a concerted effort to capture this information throughout the recent history of government initiatives. The two stools between which this implementation challenge seems to fall, the government instigators and the industry enactors, remain clearly separated, with only the motivations of the former ever being expressed. Achieving some level of clarity about the true motivations of both government and the construction industry should help to show just how far apart those stools currently are.

And should there be any doubt of the need for that missing piece to be found, this graph from a review of energy use within offices by their EPC (Energy Performance Certificate) rating shows just how easy it is to be lulled into the false belief that we're making good progress.

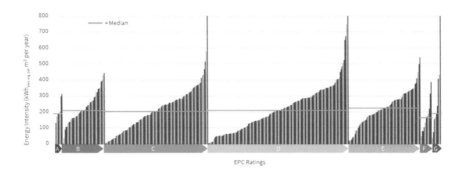

Figure 1.1 Energy use across offices with EPC ratings from G to A.

Source: Better Buildings Partnership, 2019.

Bibliography

Better Buildings Partnership. (2019). *Design for performance: A new approach to delivering energy-efficiency in UK offices*, June. www.betterbuildingspartnership. co.uk/our-priorities/design-performance

Bossom, A. C. (1934). *Building to the skies: The romance of the skyscraper*. Studio Publications.

Bourn, J. (2001). *Modernising construction*. UK: National Audit Office, January. http:// scholar.google.com/scholar?hl=en&btnG=Search&q=intitle:Modernising+Const ruction#0

Climate Assembly UK. (2020). *The path to net zero*. www.climateassembly.uk/report/ read/final-report.pdf

Eclipse Research Consultants. (2009). *Never waste a good crisis: A review of progress since Rethinking Construction and thoughts for our future*. https://constructingexcelle nce.org.uk/wolstenholme_report_oct_2009/

Farmer, M. (2016). *The Farmer Review of the UK construction labour model: Modernise or die*. www.constructionleadershipcouncil.co.uk/news/farmerreport/

Integrated Project Initiatives Ltd. (2014). *The Integrated Project Insurance (IPI) Model Project Procurement and Delivery Guidance,* July. www.gov.uk/government/publicati ons/integrated-project-insurance

Part I

What Are We Trying to Do?

What's Stopping Us? How Could We Approach It Differently?

1   Defining the Sectors

The first challenge is to fully define those different sectors, not just within the construction industry, but within the whole transactional process of constructing a building, all of whom have, or should have, a voice in any decisions taken around transforming the industry.

A top-level definition provides us with five distinct categories which we define here by the roles they play: Government Controlled; Supply Chain Businesses; Building Providers; End Users; and Designers. Designers, in itself defining a wide range of stakeholders, is shown centre stage because it collectively represents the most connected sector within the build process and therefore has the most pivotal role to play in this story.

The task now is to subdivide these until there is no further subdivision possible, the purpose being not to create as many categories as possible, but to then decide what is the smallest number of categories needed to capture all the different perspectives that we need to consider. This is important but far from straightforward, as there are multiple ways in which each of these sectors can be broken down further, such as their roles in the process; the size and scope of their operations – local to international; whether they operate in the public or private sector, etc. Only if these divisions are likely to result in different motivational drivers, however, do they need to be included.

To take government-controlled decision making as an example, there are now four government departments that have a direct influence over the construction industry – the Department for Business and Trade; the Department for Science, Innovation and Technology (born out of the now disbanded Department for Business, Energy and Industrial Strategy (BEIS); the Department for Levelling Up, Housing and Communities; and the Treasury, plus other influential public bodies such as the Climate Change Committee (CCC), now with its own new Department for Energy Security and Net Zero to harangue. To this we could add a second form of division in all the ministries that deal with specific sectors where government procures for constructing schools, prisons, hospitals, law courts, etc. Each of these have their own agendas and each will be competing for funds from the Treasury whilst keeping one eye on their own party's chances of re-election, all of which are drivers that influence the decision-making process.

DOI: 10.1201/9781003332930-3

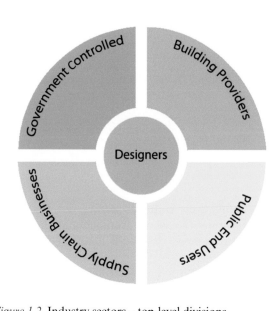

Figure 1.2 Industry sectors – top-level divisions.

The government is not an easy place to establish motive at any point in time, let alone into the future. Priorities can change on a pinhead and what is driving policy one day can be superseded by more immediate concerns the next, which is partly why the sustainability industry has struggled to establish itself on the back of global events and inconsistent changes in government incentives, tariffs and legislation. So, in order to define those sectors with similar motivations and shared perspectives, it is perhaps more beneficial to work with the overarching objectives rather than at the more changeable policy level. The CCC is clearly driven by a sustainability agenda, whereas the Treasury is driven by value for money, productivity and GDP. Beyond that, different sectors will be championing different causes, with housing being possibly the hardest to pin down with ten ministers in ten years and a constantly changing allegiance to other combined departments.

Figure 1.3 sets out the drivers that are suggested here to exist, categorised under financial, societal or environmental benefit, or in many cases a combination of any two of these three primary categories. By the end of this chapter we should be able to match these drivers to the relevant industry stakeholders we're about to define, for whom they are key decision-making factors.

Once completed, the matrix created from this information can then be used in either one of two ways. It can be used to show what the main motivating factors are behind the decision making of any specific stakeholder, or to show which stakeholders are likely to support or resist a policy based on any specific motivating factor. It's an iterative process, and at first may include some generalisations that gloss over some further subdivisions that need to

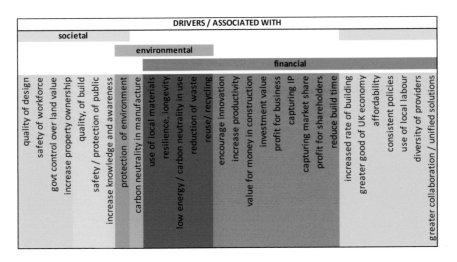

Figure 1.3 Industry drivers.

be included, but the intention is merely to use this matrix as a starting point for the discussions that need to be had to develop some more collaborative strategies around shared values, and also to recognise the many alternative viewpoints that exist that highlight the reasons behind some existing strategies' more disappointing impact levels.

Seek out the missing perspectives

Starting off this process in Part I by looking at some recent government initiatives, serves to corroborate this approach, as the evidence exists here for how those initiatives have been received. It also allows us to examine the process in more detail, first by looking at the predominantly political drivers for change around which the policies have been developed (Chapter 2), and then at the barriers confronted in terms of those other stakeholders' conflicting drivers that were never fully considered (Chapter 3). From this, a complete process for collaborative decision making emerges that guarantees the willing participation of all parties (Chapter 4) that is then applied in Part II to the various issues known to exist, but that remain as yet unresolved. But first, we need to fully define these sectors and what divides them, both internally and from each other, starting with those roles loosely defined as 'government controlled'. The process we're adopting here and for each of these sectors is to first define the many ways in which they can be subdivided and the subsectors that exist in each case, and then to assess where the motivational differences lie so that we can then group some of these together where they are similar enough to allow us to reduce the total number of subsets we need to be dealing with to as few as is required.

Don't over-complicate the problem

◆ **Government** • Contractors • Suppliers • Designers • Public End Users

Government-Controlled Roles

This is a necessarily loose definition as the perceived remit of governments changes across party lines, with Conservative governments conventionally more inclined to limit that role and consider privatising certain aspects, such as their control over the inspection of the construction process on-site. There is also a fluid division between centralised control and the devolution of some powers to the regions and city authorities, all of which sets up the possibility of conflicting motivations within this sector as a whole.

This is not a discourse about politics in the UK but merely an attempt to bring some commonly known facts into the equation so that the right questions can be asked. Perhaps more than in any other sector, the fluidity of politics and the pace at which priorities can change means that adopting a questions-based approach here will have a far longer shelf life than trying to predict or pin down intended outcomes. The fundamental question is: 'What is it that motivates any government-controlled decision making and where do the division lines lie within this sector?', remembering that the answer to this must be broad enough to capture any scenario under any party's leadership. The primary 'response divisions' have already been defined as Financial, Societal and Environmental, against which the different divisions within government need to be assessed. At one level there is central and local government, but taking central government as the model, there is the Treasury, clearly focused on the economy as a whole, Housing (currently The Department for Levelling Up, Housing and Communities (DLUHC) and other ministries focused on provision at a societal level, and some evolving sense of an environmental crisis, driven by the CCC, but necessarily influencing policy across all departments.

Focusing on the housing sector for a moment, on the surface there may appear to be a consensus around the need to provide more, better quality, truly affordable housing, but these three aims are, in reality, conflicting requirements. At an individual project level, clients are often asked where their priorities lie between speed, quality and price, since they cannot expect to achieve all three, only two. Fast and cheap will impact on quality, quality at speed will increase costs, and quality at an affordable price will take time. The same rule applies at a policy level, resulting in a conflict between the Treasury, focusing on productivity and cost, the ministry for housing, currently focusing on speed and quantity, and the environmental lobby, pushing for a better quality, more sustainable approach to construction, which, at least in the short term, comes at a price.

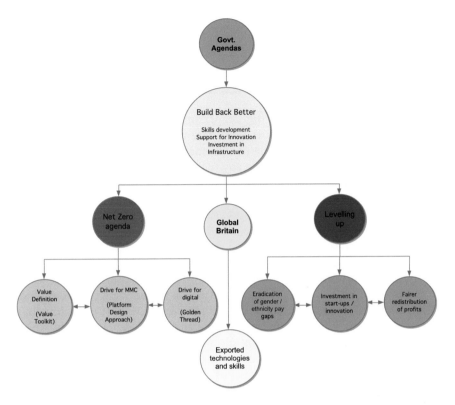

Figure 1.4 Government agendas.

Set these conflicts within the larger picture of government agendas and the complexity of the decision making that any industry lobby group would have to interact with brings home the enormity of the task at hand. Not understanding the rules of the political arena that is being entered into can result in a small return for a substantial investment in time and effort. The current initiatives attempting to transform the construction industry can be shown to draw upon all of the government's key objectives, although 'Global Britain' in this context is still largely aspirational.

Achieving an outcome that satisfies all agendas can quickly look like an almost impossible balancing act, even within the political arena, but there are still more variables in play to consider when looking at the interface between government and industry. We have established that there are conflicting motivations within government, but is it right to assume that there is a complete understanding across these departments of the industry's problems that need to be addressed? Do we know where the information being used at a governmental level to make these decisions is coming from? Is it balanced, objective and complete, or are they too working with a subset of the data needed to correctly inform their policy choices? And, equally important, to what extent can government actually influence the industry with the policy decisions they choose to make? Where does government sit in that hierarchy of influence when considered alongside the power of businesses or the general, voting public?

What is the hierarchy of influence

What becomes clear as we dig deeper into the complexities of this or any other industry's dilemmas is that there are not only many conflicting perspectives, but also a hierarchy of influence to consider before arriving at a strategy that can be said to have any realistic chance of success. And it is the impact of that hierarchy of influences on the decisions being made where the real fluidity lies, and where the balancing act that plays out between these conflicting requirements can be most easily seen. To use a stark example, the need to speed up housing delivery resulted in the Cameron government's 'bonfire of red tape', just as the death of 72 people at Grenfell has now resulted in a shift back towards stricter regulation. Similarly, the need to 'build greener' resulted in a 'presumption in favour of timber' proposal from the CCC (Holmes et al., 2019), with one authority even introducing a 'timber-first' policy (London Borough of Hackney, n.d.; Wu & Hyatt, 2016), resulting in a flurry of high-rise CLT (compressed laminated timber) developments, which has since been rescinded for reasons discussed in Part II.

Further examples of 'hierarchical fluidity' can be found between the government's objectives and those of the housebuilders, supply chain manufacturers, warranty providers, and of course the public end users. Our democratic process, sometimes described as being closer to an elective autocracy (Taylor, 2019; Walker, 2021), means that the public's demands only influence policy decisions when securing their vote becomes a priority, but even then, in this scenario that represents a balancing act between providing more housing for those not yet on the housing ladder, against limiting the rate of build to protect the value of existing homeowners' assets.

These are what are known as 'Wicked Problems' – complex societal issues that cannot be solved as such but only resolved at the time of asking, the point being that any solution reached will need constantly updating, dependent on the situation as it exists at that time (Buday, 2017). It is for this reason that establishing a series of questions that need to be asked represents a far more robust approach to problem solving at this level than trying to define fixed solutions that will inevitably pass their sell-by date very quickly.

Wicked Problems cannot be solved, only periodically (re)solved

---

In Chapter 4 these concepts are brought together to suggest how such a question-based approach could form the basis of a decision-making process that can make sense of the many conflicting interrelationships that define our construction industry. But before we move on from 'government controlled' to the other sectors, have we managed to fully define where the division lines need to be drawn between the different motivators within this sector?

The candidates for this are 'political ideologies'; 'national / local policy drivers'; 'pragmatism / idealism' (or short term / long term); and 'financial / societal / environmental'. There is a fifth motivator that is possibly too contentious or too uncomfortable to accept, but nonetheless a clear contender, and that is 'the greater good / self-interest', which can equally be applied to political parties as it can be to the individuals within them. These five divisions represent the political battleground of ideas within which decisions are made based upon a fight for motivational supremacy. There are however some correlations that can be seen to exist amongst these motivators in Figure 1.5, which can at least be used to set expectations, if not determine outcomes.

The takeaway message here is that lobbying government is a long-term game, and more of an art than a science, requiring in-depth and up-to-date knowledge of government policy and the individuals involved, alongside an awareness of how fast and superficially those far-reaching decisions are often taken (Thorpe & Andrews, 2018).

So what about the construction industry itself, the entity that government is hoping to influence with these ideas enacted as policies? The construction

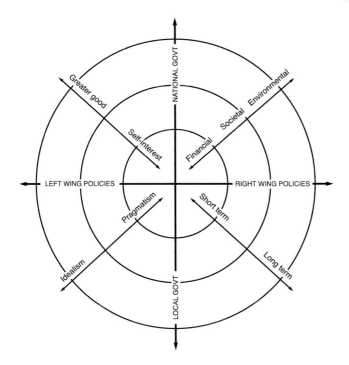

Figure 1.5 Government motivators.

industry has already been subdivided for this task, into the providers of the finished products, the suppliers of those products' component parts and services, and the designers who together define how those products will be created and used. These are task-related divisions rather than motivational, but provide a natural starting point for looking at where those drivers might differ. There is no obvious order in which to tackle these other than a perceived hierarchy of influence, which, in the case of housing, would put the housebuilders next in line if not ahead of government itself, another under-emphasised but important factor, to be revisited later.

⬤ Government • **Contractors** • Suppliers • Designers • Public End Users

Contractors / Developers

The entire construction industry is defined by BEIS as belonging to one of three sectors – Infrastructure (~15%), Residential (~40%) or Commercial and Social (~45%) (Eykelbosch, 2021). For this exercise, provision within the residential sector can be defined in any of four ways:

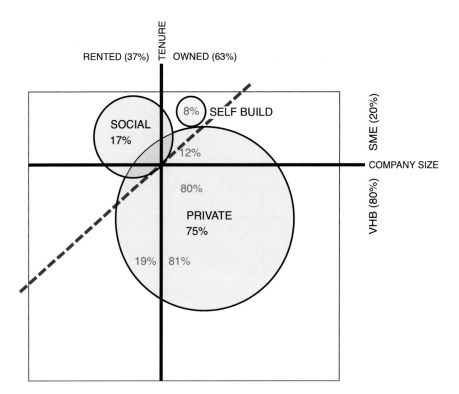

Figure 1.6 Divisions within the housing industry.

By the size of the provider organisation;
By sector – social, private or self-build;
By traditional or modern methods of construction; or
By tenure – rental or ownership.

In this instance, however, there are some strong correlations between these four definitions, with around 80% of all private development being both for sale, not rental, and built traditionally by volume housebuilders, not SMEs (small to medium enterprises). Social housing on the other hand is predominantly built by small companies in the form of local authority departments or housing associations, often using innovative and sustainably driven solutions, and for the rental market as either social or affordable rent housing.

Again, this is a fluid division, with the role of SMEs collapsing since the 2009 recession from around a third of market share to their current position, and the provision of social housing having been on a steady downward trajectory since the late 1960s (Shelter, 2022), although now through necessity beginning to rise again. What this shows is a clear division within the housing industry of motivational factors that, when combined with political ideology, begins to explain why the volume housebuilding sector sits above others in terms of its hierarchical influence: As our main housing provider, they can and do strongly influence government policy in terms of building regulations and planning requirements, and also the extent to which they agree to provide social housing and local amenities as part of their housebuilding remit, in return for which they 'endeavour' to increase their rate of supply (Archer & Cole, 2016).

This is possibly the most extreme example of motivational conflict within the construction industry – a government desperately wanting more housing to be built to ensure, in part, that more than 50% of the voting public remain homeowners, and 80% of our housebuilding provision coming from a sector whose business model requires them to build at a slow rate that sustains a high demand and guarantees maximum profits. The complexities of this dysfunctional relationship run far deeper, but that is something to unpick in detail in Part II once the tools have been developed to allow for a full investigation. As part of that investigation we will also begin to understand why only 8% of our housing is self-built, despite there being evidence of a third of our population expressing an interest in doing so (Hayes, 2020), why the SME sector has shrunk to only 12% of market share, and why our social housing provision has all but stopped, despite numerous government initiatives to sustain it (Grayston, 2021).

So, for the housing sector, what are the division lines that best define these different motivators, and to what extent can this be extrapolated for all contractors / developers? Clearly this has a lot to do with business models, with the volume housebuilders holding the trump card: With a large proportion of their profits coming from the uplift in land value realised by building out a site, and the majority of the uplift in land value being due to the length of time that site has been in their ownership, anyone entering the market and hoping to compete on price has a near impossible task on their hands. Maintaining that advantage therefore becomes a fundamental driver for this sector, as does breaking the link between build cost and land value for those competing against them (Payne et al., 2019).

Along similar lines, the volume housebuilders' business model benefits from maintaining the status quo in terms of well-established methods of construction, whilst the 'New Entrants', companies from outside the construction industry with manufacturing methodologies, are keen to employ any new approach that might disrupt the market and change the dynamic. Using increasing health and safety regulations, sustainability targets, and the demand for a faster, more productive delivery model, all of which help drive off-site manufacturing and modern methods of construction, is part of this effort to rebalance the playing field by competing under different rules. In this instance, therefore, a number of aligned factors come together to represent a major motivational divide between small, socially driven innovative businesses working predominantly in the public sector and a few large traditional speculative housebuilders providing for the majority of the country's existing and prospective home owners. As yet, government has failed to find a way to balance its own objectives and support a much needed broadening of the supply base to speed up delivery without that resulting in a crash in property prices (Gardiner, 2020), but as we will show in Part II, the solutions to this conundrum are attainable, given the will to act.

As for the broader non-residential sector, the same divide exists to some extent, reflecting the need for any young company to find its own niche by disruptively pushing the boundaries as opposed to the conventionally more protective stance of larger more established companies with embedded knowledge and the security of an existing client base. But this David and Goliath battle should not be drawn as good against evil, for the reality is far more complex, with interdependencies working across the divide and a need for all actors to play their part in bringing innovative solutions forward, as is discussed in Chapter 8.

✦ Government • Contractors • **Suppliers** • Designers • Public End Users

Material and Trade Suppliers

This sector is the most complex, but also the least visible and the least connected within the industry as a whole. The end user has little or no contact with the many supply chain companies responsible for their end product's component parts, and even the direct purchasers of those parts often fall into stunted relationships with their suppliers that limit the opportunity for any collaborative development of better solutions (Bishop et al., 2009). Governments, for their part, rely heavily on industry lobby groups to inform policy decisions, but the industry's smaller, more fragmented supply chain companies struggle to make their case with a collective voice, limiting their political influence (Bratt, 2021).

It is for this reason that many supply chain companies, most of which are locally or regionally based SMEs, subscribe to trade association or certification bodies, but here too there exists a picture of fragmented, competing organisations limiting their own impact through duplication of effort and the splitting of scarce resources (International Building Press, 2021). Despite this, however, there is a shared challenge faced by all of these supply chain businesses, which is to have a more prominent role in the design process, and not be treated as 'dumb suppliers' – their term – providing products on demand into a programme of works pre-defined from higher up the supply chain. The problem they face as a sector is that they are in competition with each other, and need to compete on either price or their product's USP (unique selling point) making collaboration a challenging concept. So for this exercise, the division lines tend to lie between the competing material or technical solutions on offer, such as different foundation solutions, concrete vs. steel frame construction, etc., where trade associations can make a case for their offering over any other.

At the root of these divisions are the base materials of construction – timber, steel, brick and block, and concrete, each of which can be subdivided into different constructional solutions that also compete against one another, such as precast and in-situ concrete; CLT, SIPs (structural insulated panels) and timber frame, etc. In a similar way, methods of delivery spawn different company solutions, beginning with on-site and off-site, and further subdivisions within each, such as panellised or volumetric construction, which then overlap with the base materials used. It is perhaps unfair to expect there not to be some inevitable crossover between trade associations' remits, but the result of this is, for instance, a timber frame company having to choose between belonging to TRADA (Timber Research and Development Association) to promote the use of timber, the STA (Structural Timber Association) to promote the use of timber frame, BOPAS (Build Offsite Property Assurance Scheme) to promote off-site manufacturing or NaCSBA (National and Custom Self Build Association) to promote the self-build industry – all of which could be beneficial, but all of which mean paying membership fees. Similar choices or dilemmas exist within other material sectors, on top of which come additional choices, such as which environmental lobby group to associate with to promote that material solution over any other.

Material suppliers represent only one part of the supply chain sector, however. The three primary subdivisions are Materials, Equipment and Trade Services.

Equipment – plant, site cabins, scaffolding, skips, etc. are either purchased or hired, representing two different business models, but the overall demand for equipment is dependent primarily on the choice of on-site or off-site methods of construction.

Trade Services captures the labour requirement, traditionally defined by 'the trades' – groundworker, erector, bricklayer, joiner, plumber, electrician, roofer, plasterer, decorator, but now becoming more fluid as MMC (Modern Methods of Construction) solutions blur the boundaries and subdivide these roles along different lines.

So these are the technical divisions that can be used to define this sector, but to what extent do they represent divisions that define similar motivational drivers, even if the communication channels between them are limited? And if there are further differences that need to be identified, what are the alternative

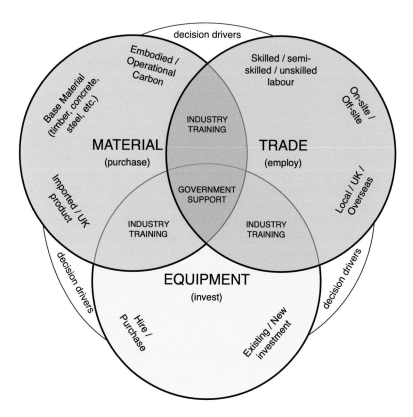

Figure 1.7 Supply chain divisions.

division lines we should be using? One way to approach this is to look for the opportunities and threats posed by possible future events, such as climate change, increasing health and safety requirements, the growth of off-site manufacturing, a declining pool of skilled labour, the rising cost of imported goods, political ideology, recession – the list is long and a valuable resource for informing decision making if used systematically as part of a 'Wicked Problem' approach. Again, this is another part of the process that will be added into the equation in Chapter 4 to develop the decision-making tool that will be used in Part II to suggest some alternative strategies.

Agenda-driven innovation

For now there are some key division lines that stand out for further sub-dividing the supply chain sector, but none that do not need more investigation before being used to define differences of purpose. One of these can be found in the environmental lobby camp, and their use of the net zero agenda to promote the use of timber over all other construction materials. In reality the benefits of timber in this respect are extremely difficult to calculate, let alone police, even by the timber industry's own admission (Morison et al., 2012), but the pressure on them to do so is negligible due to the near universal acceptance of this benefit as an undisputed fact. Many of the arguments needed before decisions are taken at such an influential level are far more nuanced than those used by powerful lobby groups to promote their own industries within the political arena. Other materials' embodied carbon calculations are not only far easier to calculate, but also far easier to reduce through technological advances in methods of production, not to mention the introduction of green energy, which in itself can completely change the figures for embodied carbon against timber's 'undisputed' advantage. And in addition to this there are other factors to consider, such as carbon costs associated with imported materials (80% of all UK construction timber), biodiversity loss and resilience to global economic factors.

Environmental impact is only one of many examples of where divisions can be drawn within this sector that represent differing motivational drivers. Increasing health and safety requirements have become a major driver for off-site manufacturing, not only because of it being a safer environment to work in but because accidents that do occur are easier to monitor and record. A more fundamental driver for off-site manufacturing, however, that transcends the compliance requirements, is the unwillingness of a depleted labour force to work in historically poor on-site conditions, usually under short-term contractual agreements.

Again, all of these are either complex arguments that need to be understood at a deeper level, or fluid factors that can change rapidly. Both require a system for fast, accurate evaluations to be made prior to strategies being developed, and even, on occasion, re-evaluated during the timescales of a project on-site.

Decision making needs to be seen as a continuous process

 Government • Contractors • Suppliers • **Designers** • Public End Users

Design Service Providers

This is where the meat of the design process lies and it is the sector most exposed, if not open, to change. In the past, and arguably up until the Thatcher era, the architect was still holding onto the role of 'Master Builder', with oversight of the entire build process from inception to handover. Those days are now long gone, primarily due to a renewed focus in the 1980s on cost and productivity and the architectural profession's innate ability to deprioritise that factor in the equation in favour of almost every other consideration. Unfortunately, with the loss of that status also went the loss of any oversight role, together with the 'fat' in the total build cost to pay for it. What we are now left with is a series of processes carried out with varying degrees of co-operation between different disciplines, all with their own motivations and cost centres to prioritise.

The divisions in this sector are therefore fairly well drawn already and can even match to some extent the linear progression that is our recognised build process – the Concept Design; Developed or Costed Design; Technical Design, including structural and building services information and any other specialist subcontractor's input; actual Construction; and Handover. Within this process there are some key professional services that fulfil these different roles regardless of any hierarchical structures that exist within or between the companies involved, and many of these professions are still perceived to operate in ways that create tensions between them. The architectural profession, possibly the most exposed and best known for its foibles, has already been condemned for tending not to understand the implications of its own design decisions, neither financially nor structurally. The quantity surveyor, certainly from the architect's perspective, knows 'the price of everything and the value of nothing', to recontextualise Oscar Wilde, and is held responsible for 'value engineering' out the design decisions that were there to lift the building above the ordinary, whilst the structural engineer is accused of calculating the necessary structural requirements to comply with the regulations and then doubling them to ensure there's never any come-back. And the building services industry for their part are seen as the 'easy way out' when design decisions were not taken early enough to negate the need for their input – not their fault exactly, but also not in their interest to design themselves out of a job.

These are often outdated perceptions, and certainly generalisations, but the underlying story is one of the perpetuation of a system that works in silos of self-interest, with little collaborative oversight looking to maximise the benefit within the build programme for the client or end user. The danger lies in the often stated belief that these design industry sectors are wilfully refusing to work collaboratively for the petty reasons outlined above, and that the problem is theirs alone to rectify. What we intend to uncover here are the more endemic problems that stand in the way of true collaborative working and suggest how they could be remedied to the satisfaction of all the parties that would need to be engaged with that transition for it to be successful.

The starting point is therefore to fully understand the divisions that exist across these roles. The end point, to keep it in sight, is to recognise all the mutually beneficial ways in which working collaboratively benefits not only the client, but each of the roles currently being carried out sequentially rather than as an iterative process. The progression through these roles is not always carried out in the same order or originated at the same start point, and is influenced by the main drivers behind any project, previously described as being a combination of cost, speed and quality. Quality might result in the architect being invited to make the first move, whereas a project driven by cost or speed might find the architect being consulted at a later stage, or at least given a reduced remit.

In an ideal world, of course, all of these disciplines should be working together from the outset, informing every decision taken, and continuing that engagement throughout the design and build programme, which is where the technical barriers that exist begin to expose the extent of the problem to be overcome. BIM (Building Information Modelling) is both an essential component in the sharing of information backwards and forwards amongst the design team, and also the catalyst for breaking down the barriers that exist in the first place so that the whole design team can begin to understand the design process from each other's perspectives. Our painfully slow roll-out of BIM (Royal Institution of Chartered Surveyors, 2020), it will soon become clear, is the sticking point and the problem most in need of our attention, something that will be revisited in Part II.

Figure 1.8 attempts to show how these too loosely connected disciplines currently operate, and which of the three fundamental drivers, speed, quality and cost, has the most influence on their decision-making processes.

Whether or not the architect always has an understanding or an influence over all these areas is debatable, but the fact remains that no other party has that overall interest or receives as much training outside their own area of expertise, and all other sectors are therefore limited in their understanding of how those other parties operate or reach the decisions they do. As an example, the operational cost of a building is of no direct consequence to anyone other than the client, and therefore the architect and the environmental consultant charged with reducing the building's overall environmental impact are left to fight this corner alone, which goes some way to explaining why it is that environmentally driven solutions are so difficult to embed into the build programme. But if those solutions carried a cost advantage, the financier, the quantity surveyor and the contractor would be aligned behind them and the outcome would be

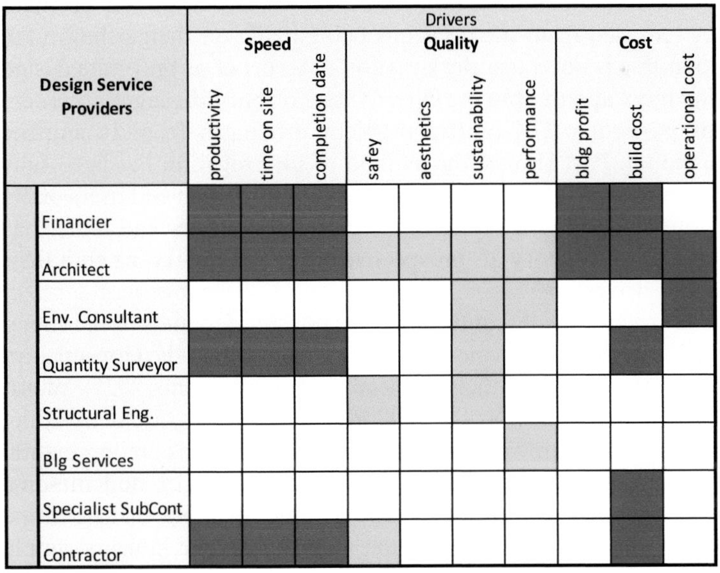

Design Service Providers	Drivers									
	Speed			Quality				Cost		
	productivity	time on site	completion date	safey	aesthetics	sustainability	performance	bldg profit	build cost	operational cost
Financier	■	■	■					■	■	
Architect	■	■	■					■	■	■
Env. Consultant										■
Quantity Surveyor	■	■							■	
Structural Eng.										
Blg Services						■				
Specialist SubCont									■	
Contractor	■	■	■	■						

Figure 1.8 Drivers within the design sector.

very different. Defining a cost advantage is therefore a key element in bringing about change, and something that can on occasion require no more than the translation of the same message into a more understandable language.

Define the direct cost benefit for all stakeholders

Government • Contractors • Suppliers • Designers • **Public End Users**

Public End Users

One of the main reasons for carrying out this exercise is to reconnect the build process with the end users, who have become increasingly distanced from the products that they are, either directly or indirectly, paying for. There is perhaps no other area of commerce where the client has so little influence over the item being purchased, with the residential market being the worst offender. For the purposes of this exercise, the division of residential and non-residential is a useful starting point, as the subdivisions within each of those categories follow different paths. Non-residential consists of commercial, retail, leisure, educational, care, health, accommodation, and justice, whereas residential is better categorised by tenure: owner-occupier, renter, prospective buyer – and then by method of procurement: private, social or self-built. Possibly the main division of interest lies in who is the owner with an ongoing responsibility for the building in terms of maintenance and operational cost. Anything built speculatively, either residential or non-residential, means the final occupier is even further removed from the decisions being made on their behalf, with the evidence that this is not a healthy situation in terms of appropriate design outcomes being most apparent in the current state of our housing market.

Where that particular UK-centric problem originates from is another area for discussion in Part II once the method of interrogation has been fully evolved, but for now the task of defining the divisions between end users offers complexity enough. Within the residential sector, home owner and renter has to be joined by a third category of 'prospective buyer', as they come with their own unique set of drivers.

Governments are continually conflicted by the housing demands of the voting public, even before having to placate the industry from which the Conservative Party at least derives over 20% of their financial support (Williams, 2021). Whilst prospective buyers, also known as future Conservative voters, are demanding more housing at affordable prices, existing home owners, seen as predominantly Conservative voters, have become increasingly accustomed to, and in some cases dependent upon, their property's growing value. Walking that tightrope would be difficult enough, even without the majority of our housing supply being controlled by a handful of developers whose business model benefits from them undersupplying the market, but with that factor more or less dictating the

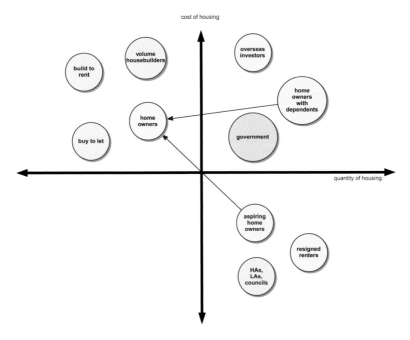

Figure 1.9 Drivers amongst end users.

options available to government, the slow but inevitable slide towards minority home ownership seems to be proving difficult to arrest (Storey & Coombs, 2021). This constitutes a classic Systems Thinking dilemma, as those very same prospective buyers, once on the housing ladder, have a tendency to 'swap camps' and become part of the resistance movement against more housebuilding in order to protect their own investment, until that is, their children are ready to leave home.

The solution to this runs far deeper than the froth of normal political debate around the subject, but for now the camps can be adequately defined if we include the fluid state of residential ownership and the influence that has on the drivers that sit behind the decisions taken. For non-residential, it would seem that tenure is also a more important factor than the subdivisions defined here, but these are only best guesses at this stage – and this, merely to define the categories, not the thoughts that lie within them. The trap of assuming the motivations of other sectors, especially the general public – who are effectively 100% of the population – is a common failing of the housing industry in particular, and patently absurd. There are a raft of further divisions that can be defined relating to demographics, psychographics and sociographics that first need to be studied before any attempt can be made to second guess the needs and motivations of an entire population when making what is, for the vast majority of us, the most expensive purchase of our lives.

residential	owner	self build
		purchased
		prospective
	renter	private sector
		public sector
		social housing
non residential (owned or rented)	accommodation	student
		elderly care
		hospital
		prison
		hotel
	retail	shops
		markets
	commercial	offices
		factories
		warehouses
	leisure	sports
		entertainment
	educational	nursery
		school / college
		university

Figure 1.10 Divisions within residential and non-residential.

◈ **Government • Contractors • Suppliers • Designers • Public End Users**

Making Sense of the Divisions

The methods used to subdivide these five main sectors that together capture all those parties that either do or should have an influence over the construction industry's decision-making process vary considerably. There is no pattern to this, which reflects the complexity of the divisions that need to be better understood before developing and implementing change initiatives. The approach taken here was to start with what was either known or suspected and to then interrogate those assumptions to establish whether or not they held up. As suggested earlier, the initial phase of this process often results in more and more subdivisions being uncovered before then collating these into a smaller number of 'close fit' categories that define starkly different motivational drivers. Where the genuine opportunities for collaboration between sectors exist then becomes clearer to see, alongside where there are differences of opinion that need to be worked around rather than run in to.

Barriers to change can be avoided as well as removed

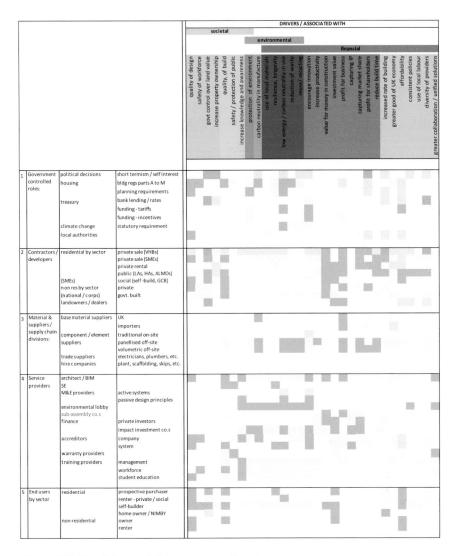

Figure 1.11 Breakdown of drivers across all sectors.

The table of drivers created earlier, grouped under Financial, Societal and Environmental, has now been used to create a matrix based on the subdivisions defined for each sector to start that process. For clarity, these drivers have all been defined as positives in terms of their broader benefit, so innovation appears as 'encouraging innovation', which is a driver for many, as opposed to 'resisting innovation', which may also be a driver, but only for those with a financial interest in protecting the status quo. In these instances the result is marked in red as opposed to green, and where there is little or no direct relevance, left as blank. The lighter shades of red and green represent a more passive impact rather than an attempt to influence outcomes, as in, for example, the planning system's impact on the rate of building – a negative but not its primary intention. And these are measures of actual outcome, not intended outcome – the volumetric housing sector may well intend to boost productivity rates, but as yet there is no evidence of this becoming a reality. What this all reveals is a patchwork of different and, at times, conflicting motivations that may begin to explain some of the more deep-seated barriers to change within this complex industry.

The evidence for some of these entries has yet to be discussed, and the picture is in no way complete. Neither is it resistant to change as events occur that inevitably change priorities, throw up new solutions, and reset trajectories at a personal, business or political level. But before trying to decipher the patterns of behaviour within this matrix, we need a tool that can help make sense of so much data. And before we can build that tool, we need a better understanding of what drives change within these sectors and what stands in the way of those changes being adopted within them.

Bibliography

Archer, T., & Cole, I. (2016). Profits before volume? Major housebuilders and the crisis of housing supply. Centre for Regional Economic and Social Research, Sheffield Hallam University, 1–28. www.shu.ac.uk/centre-regional-economic-social-research/projects/all-projects/profits-before-volume-major-housebuilders-and-the-crisis-of-housing-supply

Bishop, D., Felstead, A., Fuller, A., Jewson, N., Unwin, L., & Kakavelakis, K. (2009). Constructing learning: Adversarial and collaborative working in the British construction industry. *Journal of Education and Work*, *22*(4), 243–260. https://doi.org/10.1080/13639080903290355

Bratt, S. (2021). To be heard, we must speak with one voice. *Construction News*, April. www.constructionnews.co.uk/skills/to-be-heard-we-must-speak-with-one-voice-09-04-2021/

Buday, B. R. (2017). *The confused and impoverished state of architectural research*, 1–9. Common Edge. http://commonedge.org/the-confused-and-impoverished-state-of-architectural-research/

Eykelbosch, J. (2021). Construction statistics, Great Britain. *Office for National Statistics*, *000*(October), 1–13. www.ons.gov.uk/businessindustryandtrade/constructionindustry/articles/constructionstatistics/2020

Gardiner, J. (2020). Jenrick plans to break dominance of volume housebuilders. *Building Design*, September. www.bdonline.co.uk/news/jenrick-plans-to-break-dominance-of-volume-housebuilders/5108072.article

Grayston, R. (2021). Squeezed out: The impact of build costs and planning reform on social housing supply. *New Economics Foundation*, *44*, 1055254.

Hayes, D. (2020). *Self build aspirations*. NaCSBA.

Holmes, G., Hay, R., Davies, E., Hill, J., Barrett, J., Style, D., Vause, E., Brown, K., Gault, A., Stark, C., Harry, S., Scudo, A., Budden, P., Beasley, G., Labuschagne,

C., Freeman, B., & Darke, J. (2019). *UK housing: Fit for the future?*. Committee on Climate Change, February.

International Building Press. (2021). *Trade associations and industry groups*. https://ibp. org.uk/trade-associations-and-industry-groups/

London Borough of Hackney. (n.d.). *London Borough of Hackney Sustainable Design and Construction Supplementary Planning Document*. https://hackney.moderngov. co.uk/documents/s43377/LHR%20K31%20Appendix%201_SPD_June_Cabinet_2 015.pdf

Morison, J., Matthews, R., Miller, G., Perks, M., Randle, T., Vanguelova, E., White, M., & Yamulki, S. (2012). Understanding the carbon and greenhouse gas balance of forests in Britain. *Forestry Commission*. https://cdn.forestresearch.gov.uk/2012/05/ fcrp018.pdf

Payne, S., Serin, B., James, G., & Adam, D. (2019). How does the land supply system affect the business of UK speculative housebuilding? *UK Collaborative Centre For Housing Evidence*, February.

Royal Institution of Chartered Surveyors. (2020). The future of BIM: Digital transformation in the UK construction and infrastructure. *RICS Insight Article*, July, *52*.

Shelter. (2022). *The story of social housing*. Shelter, 263710.

Storey, A., Coombs, N., & Giddings, R. (2021). *Living longer: Changes in housing tenure over time*. Census 2021, Office for National Statistics.

Taylor, R. (2019). *We are still perilously close to Hailsham's 'elective dictatorship'*. London School of Economics, September. https://blogs.lse.ac.uk/brexit/2019/09/30/ we-are-closer-than-ever-to-hailshams-elective-dictatorship/

Thorpe, K., & Andrews, E. (2018). *Policy engagement*. Institute for Government, May.

Walker, T. (2021). *Ken Clarke: 'We're now close to an elected dictatorship under Boris Johnson*. New European, November. www.theneweuropean.co.uk/tim-walker-intervi ews-ken-clarke/

Williams, M. (2021). 20% of Tory donations come from property tycoons. *Open Democracy*, *20*, July. www.opendemocracy.net/en/dark-money-investigations/

Wu, W., & Hyatt, B. (2016). Experiential and project-based learning in BIM for sustainable living with tiny solar houses. In Chong, O., Parrish, K., Tang, P., Grau, D., & Chang, J. (Eds.), *ICSDEC 2016 – Integrating Data Science, Construction and Sustainability*, *145*, 579–586. https://doi.org/10.1016/j.proeng.2016.04.047

2 Drivers for Change

The Government Perspective on What the Industry Needs to Do

Our main driver for change ought to be our climate emergency, but

Priority = Importance x Imminence

which means that, in reality, climate change rarely tops the priority list in all but name, and even more worryingly, as the repercussions of climate change become more pronounced, attention will inevitably get increasingly diverted towards dealing with its consequences rather than the more intractable root causes. This is just another part of the complex 'causes and consequences' hierarchical structure that needs to be drawn in detail before attempting to suggest any solutions to the construction industry's more immediate ills, and it needs to be drawn from all relevant perspectives.

So – from the government's perspective, whilst its ambition to achieve carbon neutrality by 2050 is actually a statutory requirement, there are some far more immediate challenges to contend with, some of which could quite legitimately be defined as prerequisites to achieving this: Productivity is the government's main economic bugbear, especially with respect to the construction industry, which has been effectively flatlining for two decades now (Martin, 2021). Without increasing the speed and efficiency of how we build, we cannot reduce the cost of building, and if we cannot do that, we will struggle to afford the additional cost of building to a zero carbon agenda – or so the justification goes. It is certainly true that the zero carbon agenda has repeatedly stalled at the funded 'demonstrator site' phase, where technically proven solutions have then failed to become mainstream solutions due to the additional upfront costs involved in implementing them. It is fair to say, however, that, viability aside, what the zero carbon agenda does require us to do is to build differently, and break with our traditional use of materials and methods of construction.

If we accept 'building differently' as a loosely defined objective, the question turns again to how government can drive innovation in an industry that has proven itself to be hard to reach, let alone influence. As a question, this needs to sit above the hierarchy of drivers already established, but in reality, the reasons behind our current lack of progress are rarely discussed. Is our climate change

DOI: 10.1201/9781003332930-4

emergency really not seen by the construction industry as important enough, or the effort needed to increase the industry's productivity to a competitive level just too difficult a challenge, or is there something else at play lying behind this perceived industry-wide 'complacency' that we should be focusing on? Are the new ways of working being prescribed through government initiatives just too big an ask, or is the act of transitioning itself the real barrier to change that we need to be focusing on?

This is where the government is still working with too many unknowns, with a fraction of the perspectives it needs to shed light on the truth, and using construction-level methodologies that fall far short of the complexity of the problems that need to be solved.

We cannot solve our problems with the same thinking we used when we created them.

(Albert Einstein)

This is surely true of the construction industry, where the problems being confronted are as much societal as they are technical, but where the thinking and the solutions being applied are limited by a narrow definition of the industry that does not encompass the broader influences at play. This then, is the evidence for how our current approach to problem solving is falling short of what is needed to bring about any meaningful change in construction, or anything else for that matter.

Look beyond the problem for the solution

The Government's Perspective

There is no shortage of government initiatives to look at to establish this pattern of impact falling short of expectation, with over 90 key reports relevant to the construction industry published since the Second World War (Designing Buildings, 2023). Government white papers are the documents most often used to determine the intent if not the final form of these initiatives, as they still have to pass through both Houses before any of their recommendations might become enforced as Acts of Parliament. It is at the white paper stage, if not before, however, that the findings from the research undertaken by government departments or employed external consultancies get published and read by industry stakeholders.

The recommendations made are invariably a combination of how the industry needs to change and what government needs to do in terms of legislation to support those changes. The immediate consequence of this is a period of uncertainty whilst the industry waits to see how many of the report's recommendations are enacted in legislation by government, during which time little can realistically

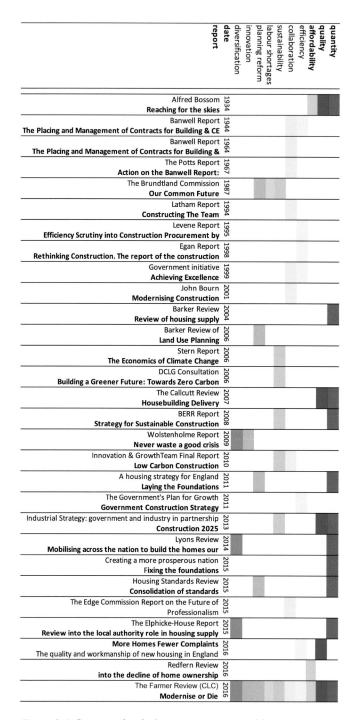

Figure 2.1 Construction industry government white papers up to the Farmer Review.

happen. This delay can often extend into the next parliamentary session and beyond, by which time the report, if read at all, has lost its immediacy and urgency as a call to action. It is not uncommon therefore to see supplementary reports being commissioned and published some years later commenting on the lack of progress made since the initial report's recommendations were passed as Acts of Parliament. But there is still little effort made to fully understand what lies behind this inaction. Is it the result of a loss of focus due to the delayed arrival of the powers needed to support the actions being recommended, or is it the failure of these reports to recognise that recommendations alone do not result in the automatic adoption of the actions being promoted? Even without 'events' further complicating the scenario, with Brexit, a global pandemic and war in Europe each being exceptional in their own right, let alone in quick succession, these are not straightforward questions to answer.

The Farmer Review is not the latest government initiative that we could use, but it does pre-date the repercussions from these events, and is for many reasons an appropriate starting point for progressing this line of enquiry: It was a broad review of the construction industry, commissioned by the government, carried out by CAST, an independent consultant, and followed up a year later by a government response in the form of the then Department for Communities and Local Government (DCLG) white paper 'Fixing Our Broken Housing Market' (DCLG, 2017) and again the following year by warranty insurance company BLP, with their own white paper 'The Construction Industry 18 Months On from the Farmer Review' (BLP Insurance, 2018). In addition, specific challenges highlighted in the report, such as the slow build-out rate on large housing sites, were tackled in separate government reports ('Independent Review of Build Out', Letwin, 2018). The real political outcome from all this analysis, however, was captured in the government's long promised Industrial Strategy (HM Government, 2017a), and as part of that, the Construction Sector Deal (HM Government, 2017b), also announced in the same year.

The Farmer Review

Everything about this review of the construction industry would suggest there was a full understanding of how change needed to be approached as a consultative process, and how the problems discovered to exist needed to be tackled holistically. Mark Farmer's report set out the shortcomings of the sector's labour model and prevailing business model, and in particular its workforce attrition, exacerbated by an ageing workforce and then Brexit. Its recommendations were focused on finding and unlocking the drivers for change, and crucially recognised that '*the most important and effective drivers for change do not necessarily sit within the industry itself*' (Farmer, 2016, p. 4).

The review highlighted ten symptoms, and from these diagnosed three 'root causes', one of which was the lack of any 'strategic incentive or implementation framework in place to overcome the issues, and initiate largescale transformational change' (Farmer, 2016, p. 8). Farmer also focused on the need for that implementation framework to involve all stakeholders:

> *Critically, a plan for change needs to recognise, based on past evidence, that the industry will not change itself unilaterally at scale. It needs to be led by clients expressly changing their needs and commissioning behaviours or government acting in a regulatory or strategic initiation capacity to drive positive disruption* … [and for that plan for change to be] *supported by a sound business case; not just optimistic ambitions, target setting or aspirational statements of intent.*
> (Farmer, 2016, p. 10)

As with all reports recommending a course of action, it warned against partial adoption:

> *It is important to note that the impact of this review's recommendations will only be maximised if adopted holistically. 'Cherry picking' will dilute effectiveness and in some instances may have negative impacts unless coordinated with other recommended measures being adopted.*
> (Farmer, 2016, p. 70)

But what of the review's actual recommendations, and how far did they go in terms of considering how they should be implemented – beyond suggesting what government should legislate for or directly finance? ('*Incentives to change should leverage as many existing fiscal or policy tools as possible*', p. 51). It began by stating that '*The case for change is compelling and industry needs to modernise itself to become a more compelling proposition for prospective new entrants or face a future of decline and marginalisation*' (p. 50). This it said should be achieved through harnessing new technologies in order to improve the industry's productivity levels, whilst recognising that this would only happen if demanded by the industry's clients or enforced through regulation. Leading on from that comes the observation that '*Interventions need to be largely capable of cross party political support*' so that '*the required shape and size of an evolving labour supply, its training and the adoption of new delivery technologies can be mapped and developed*' (Farmer, 2016, p. 51).

So far that constitutes a direction of travel for the industry and a list of government policy recommendations based upon an analysis of the industry itself being incapable of change solely from within. The vehicle recommended in the review for taking ownership of that change agenda is the Construction Leadership Council (CLC), but the following statement is perhaps the most important observation that we will be looking for action on in the subsequent white papers and industry reports:

> *Whilst the CLC can guide at a strategic level and report and make recommendations to Ministers, it is not a delivery body equipped to drive change on the ground. There is a need in any possible emerging integrated leadership model for sufficient resource and focus on how the implementation road maps being developed by CLC can be connected to industry at scale. CLC will need to create followship as far as possible across the supply chain, all types of private clients and government. This can only really be achieved by suitable balanced representation around the table.*
>
> (Farmer, 2016, p. 53)

This is the missing piece in the jigsaw, clearly referenced here as beyond the scope of this, or in fact any, report of this nature, and central to the argument being made here – that the 'balanced representation around the table' is not happening, and that, without that process, the successful implementation of any recommendations is unlikely to occur. Furthermore, that balanced representation is not just something that needs to be brought into play to understand how to implement policy decisions already made, but something that must be integral to the formation of those policy decisions in the first place if they are to be deemed worthy of implementing.

This is not a failing of the Farmer Review so much as of the brief given for the consultation process, together with the way in which the review's findings have then been taken forwards. Given a broader scope, the Farmer Review's findings, and its stated root causes in particular, would have no doubt reached some more fundamentally challenging conclusions. The statement that '*the industry has evolved a "survivalist" shape, structure and set of commercial behaviours in reaction to the environment in which it operates*' (Farmer, 2016, p. 8) is patently not a root cause of the problems being confronted, but a symptom of a deeper malaise. It is within 'the environment in which it operates' that the real root causes of the industry's ills will be found, and that environment cannot be fully understood without that balanced representation around the table from the outset.

Digging deeper to find the real root causes is for Part II, as it will involve the use of the more holistic analytical approach being developed, but for now we need to follow the path of the recommendations made and see how they have been enacted by the government, the industry and the CLC in their nominated role as overseers:

> *The Construction Leadership Council (CLC) should have strategic oversight of the implementation of these recommendations and evolve itself appropriately to coordinate and drive the process of delivering the required industry change programme set out in this review.*
>
> (Farmer, 2016, p. 53)

Recommendations aside, however, we first need to deal with the government's own response to the Farmer Review to get a better understanding of their perspective on what both they and the construction industry needs to do.

Fixing Our Broken Housing Market

As its name suggests, this report is solely focused on the housing sector, which, as the Farmer Review has already outlined, has its own specific problems to contend with over and above those more industry-wide issues, making this an even richer environment for looking at the need to consider alternative perspectives before reaching for solutions. The foundations for how the government intends to fix the problems now known to exist are summed up in the following statement:

> *This White Paper sets out the support the Government will provide to enhance the capacity of local authorities and industry to build the new homes this country needs. In return we expect professions and institutions to play their part and turn these proposals into reality.*

(DCLG, 2017, p. 16)

The government support being offered here is spread across all stakeholders – local authorities, housing associations, private developers, local communities, lenders, institutional investors and capital market participants, utility companies and infrastructure providers, each of which are then given the roles they are expected to play in return. The support being offered as part of this deal comes in the form of incentives in return for compliance with regulatory requirements.

So far, it would seem there is already an understanding that the problems being confronted require broad holistic solutions that need to be co-ordinated across all stakeholders, involving a combination of both carrots and sticks to be effective. This is, effectively, the 'industry change programme' that Farmer defined as essential for the industry to survive, and contains the detail needed for the CLC to pick up on and coordinate in order to drive the process of delivery forwards. So what exactly are the deals being offered to each of these stakeholders, and are they realistic enough for those stakeholders to engage with?

Local Authorities

The Carrots

New capacity funding;
Make it easier for local authorities (LAs) to take action against those who do not build out a site once permissions have been granted.

The Sticks

All LAs should develop an up-to-date plan with their communities that meets their housing requirements (since recinded, soon to be re-established?);

LAs to decide applications for development promptly and ensure the homes
   they have planned for are built out on time;
Central government intervention should these conditions not be met.

Comments

The deal here is for LAs to be given greater powers and new funding in order
to remove two known bottlenecks – the slow granting of planning permissions
and the slow build-out rates after that planning has been granted. The known
issues around this are that LAs have legitimate doubts that the funding may not
materialise or be adequate, and that the reasons behind the slow build-out rate
may extend beyond the scope of what the LAs can be expected to influence.
Either of these more fundamental issues being true would result in no change.

Developers

The Carrots

Faster, simpler planning system;
Improved approach for developer contributions;
Support for MMC;
Greater diversity through more support for SME builders.

The Sticks

Expected to build more homes;
Build homes swiftly where permission is granted;
Take responsibility for investing in their research and skills base.

Comments

Much of this is based on the need for the volume housebuilders to build faster
and the SME builders to build more, but does so without addressing the fundamental reasons why smaller building companies continue to lose sites to the
larger developers who will continue to build slowly in accordance with their
business model – a major topic for discussion in Part II. The demand for a
simplified planning system is driven by the volume housebuilders' desire to
increase profit margins, not build more houses, something that successive
housing ministers have failed to either understand or acknowledge.

Local Communities

The Carrot

A simpler and clearer planning process that makes it easier to get involved and
shape plans for areas at a local community level.

The Stick

Local communities have to accept that more housing is needed if future generations are to have the homes they need at a price they can afford.

Comments

The real driver for NIMBYism (Not In My Back Yard), which is not addressed in this proposal, is not so much the excessive development, but the lack of benefit to local communities in accepting that development, due to the way in which developers are 'allowed' to focus primarily on housing in favour of additional amenities – despite attempts by 106 agreements and CILs (Community Infrastructure Levies) to address this known issue (Cant, 2019). This comes back to the hierarchy of influence discussed earlier and the power that the volume housebuilders have over government decision making.

Housing Associations

The Carrot

More funding through the Affordable Homes Programme.

The Stick

Expected to build significantly more affordable homes.

Comments

Again, access to funding is not the only barrier, and this alone cannot solve the problem of delivery being largely linked to volume housebuilders' build rates, who then use financial viability assessments to create loopholes for them to escape their affordable housing obligations (Grayston, 2017).

Utility Companies and Infrastructure Providers

The Carrots

Clearer guidance;
Exploring an improved approach to developer contributions to help pay for new infrastructure.

The Sticks

The government expects infrastructure providers to deliver the infrastructure that new housing needs in good time so that development is not delayed.

Comments

This and many of the other problems being confronted here are longstanding and entrenched. Where the finance comes from is not of immediate concern to the utility companies, as their involvement is unavoidable. The drag on productivity that needs to be addressed comes from the lack of early engagement, but this requires a far more collaborative approach to the design process and contractual agreements, something that is not recognised here or throughout this paper's response.

The overall impression given by this and most government responses to the reports they commission is one of a better grasp by individual ministers of the complexity of the problems being addressed than previously existed, but still that sense of 'we've done our bit, and now it's up to you', without there being much sign of any true concern for what happens next. The reality is that with many of these proposals, the next barrier in the road is rarely considered and often remains firmly in place, meaning little progress can be made. But maybe the CLC are there to take up the baton and interpret these proposals to ensure they can be acted upon. To this end, this white paper includes a brief commitment to work '*with the Construction Leadership Council, to challenge house builders and other construction companies to deliver their part of the bargain*' (DCLG, 2017, p. 49).

The Construction Leadership Council

The CLC has been in existence since 2012 as an industry and government forum, tasked with leading progress towards the goals set out in the government's Industrial Strategy report, Construction 2025. It has undergone numerous restructurings since then in a bid to become more effective and business focused, and again recently to better represent the industry's supply chain companies. Its ability as a forum of industry experts to capture the essence of what needs to be done, however, is wholly dependent on the goodwill of enough industry representatives across the industry to give their time freely and share their knowledge openly. The reality of such voluntary organisations, unfortunately, is a lack of continuity, pace or balanced opinion. The recent observation that there is a lack of representation from the industry's supply chain companies (Mitchell, 2022) is symptomatic of this structure, where the smaller SMEs that typify the supply chain do not have the resources to commit to help develop these bodies' workstream initiatives, leading to them dropping out of the process and their needs consequently being marginalised. Without addressing the reasons behind this sector's inability to invest their time, it is unlikely the situation will improve just for having been identified.

The CLC has recently produced a publication, *CLC Strategy 2021*, in which they briefly outline their remit for the next three years. It is a short and surprisingly incoherent piece that makes no mention of the role set out for them in the Farmer Review (*Recommendation 1: The Construction Leadership*

Council (CLC) should have strategic oversight of the implementation of these recommendations and evolve itself appropriately to coordinate and drive the process of delivering the required industry change programme set out in this review), and makes little attempt to relate its objectives to that review's comprehensively analysed findings. The task of implementing the Farmer Review's recommendations 'at the coalface' still seems a long way off being recognised as a need, let alone being acted upon.

BLP – 18 Months On

So what does Mark Farmer make of the progress made since the publication of his own report? In this paper released through BLP Insurance in 2018 (BLP Insurance, 2018), he recognises that some progress has been made but the pace of change is too slow, recognising that to some extent Brexit can be blamed for that, if only due to the lack of bandwidth left to deal with this and any other issue needing the government's attention. He does however make the point that government has woken up to its responsibilities in creating an environment where change is easier for the construction industry to envisage – encouraging a broader delivery base, the use of MMC, investment and training in new technologies, a broader definition of value in its own procurement processes – all these initiatives go towards removing some of the structural barriers that continue to make change such a high-risk option. But this is still at the stage of convincing the early adopters. As Farmer admits, '*the government needs to do more to open this market up and make MMC mainstream*' (BLP Insurance, 2018, p. 4), and there is a well-known model that explains why bridging that gap is so difficult to achieve (Figure 2.2) (Moore & McKenna, 1999).

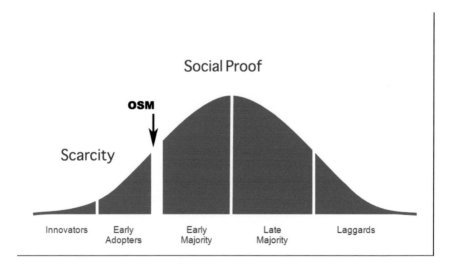

Figure 2.2 The adoption of OSM. Adapted from *Crossing the Chasm.*

Source: Moore & McKenna, 1999.

Early adopters have nothing to gain in promoting what makes their product or approach different to their mainstream competitors', meaning that getting the majority of the market to 'cross the chasm' and adopt a new way of thinking remains the job of government and the innovators themselves. Not only that, it requires a very different approach to messaging, with 'new, innovative and exciting' being replaced with 'safe, viable and proven' as the key words to ensure engagement.

Different messages for different sectors

To appeal to the mainstream markets, the underlying question to be asked is still 'what's in it for me?' but the answer now has to change to reflect the businesses asking that question – and without a full understanding of their fears and motivations, the message runs the risk of not getting through. The carrot and stick approach adopted by the government's white paper response to the Farmer Review was an attempt to grow that transition into the mainstream market, but through coercion techniques alone. It has been shown that both coercion and education are needed in equal measure to get results – neither on their own are effective (Madrian, 2014), but education is a two-way street, and begins with government better understanding the industry it wants to see transition.

Change requires both coercion and education

This final example of how government-centric the policies being put forward have become is from another government white paper that followed on from 'Fixing Our Broken Housing Market', and focused on the problems relating to the slow build-out rates being achieved after planning permissions had been granted – basically a response to the removal of one barrier, only to then run into the next. It again shows how multi-faceted, complex problems are being dealt with from the government's perspective of what needs to change to suit its own agenda, rather than that of the industry as a whole, as promoted in the Farmer Review.

Letwin – Build-Out Rate, 2018

As the then Housing Minister, Oliver Letwin's report presented recommendations for ways in which the government could increase the variety and differentiation of what was being offered on larger housing sites in order to raise the proportion of affordable housing and raise the 'rate of build out', which had been recognised as a major bottleneck in the supply process. The report rightly concluded that the homogeneity of the types and tenures of the homes on offer on these sites, and the limit to the rate at which the market could absorb such homogenous products, was key to the slow-build out rates being seen, and set about introducing ways in which these sites could in future be subdivided between multiple providers – unless a variety of house types, sizes and tenures could be provided by a single developer. The analysis behind this recommendation had found that the median build-out period on these large sites was currently 15.5 years, which it was suggested was due to 'land banking and intentional delay' (Letwin, 2018, p. 8). The solution, however, was not to address this head on, but to refer the problem to the planning authorities to deal with through their existing section 106 powers, despite recognising the need to not '*impose undue pressure on local authorities whose planning departments are already under considerable strain*' (p. 10).

The explanation for this policy decision, in Oliver Letwin's own words, goes as follows:

> *I also concluded that it would not be sensible to attempt to solve the problem of market absorption rates by forcing the major house builders to reduce the prices at which they sell their current, relatively homogenous products. This would, in my view, create very serious problems not only for the major house builders but also, potentially, for prices and financing in the housing market, and hence for the economy as a whole.*
>
> (Letwin, 2018, p. 8)

At this point, the process and therefore the conclusions reached can be seen to be limited, if not flawed, due to the single-point perspective being taken. The initial brief dictated the extent to which the report was unable to capture the perspectives needed to reach a truly beneficial conclusion. That brief was to '*explain the significant gap between housing completions and the amount of land allocated or permissioned in areas of high housing demand, and make recommendations for closing it*' and to determine '*which factors would be most likely to increase the rate of build out on these sites without having other, untoward effects*' (Letwin, 2018, p. 8). To be clear, the 'untoward effects' would be a reduction in existing house prices, unpopular amongst the government's home-owning voter base, or the disincentivisation of the volume house-building sector on which the government has become overly dependent in many ways.

An additional recommendation made was in response to the finding, again quite rightly, that the slow build-out rate was also due to a shortage of bricklayers. The error in the analysis of both this and the impact of so much homogenous housing on individual sites, however, was that they were both seen as '*fundamental drivers of the slow rate of build out*' (Letwin, 2018, p. 6), whereas they were and still are merely symptomatic of the volume housebuilders' widely understood business model. That model requires a measured build-out rate to match demand, and the building of the single house type that will return the greatest profit – which is usually three- or four-bed homes, and never affordable housing.

The solutions reached in the report were for a '*five year "flash" programme of on-the-job training*' (Letwin, 2018, p. 9) for bricklayers, and for major developers to develop larger sites with a greater variety of housing types, tenures and sizes. As one of the most profitable sectors in the construction industry, if the volume housebuilders really wanted more bricklayers, they would train their own, and if they saw a financial benefit in building a greater variety of housing, they would do so. These proposals are little more than recognitions of the problems being faced, and are unlikely to have any mean-ingful impact on the root causes that need to be addressed, but they are also unlikely to have any untoward effects with respect to the major house builders, existing house prices, financing in the housing market or the economy as a whole – as intended.

In Part II, this conundrum is revisited, taking a multi-perspective approach to show how the conclusions reached can be very different once all stakeholders' demands are entered into the equation, rather than just those of the govern-ment with an eye on the broader economy and its reliance on a sustained growth in property values. Were this just an exercise in shifting the benefit from one sector to another, the incentive to read on would quite rightly be patchy, but that is not the outcome sought or achieved. The approach being constructed is one that improves the benefit for all parties, because it is only by considering all parties' needs that any change can expect to succeed.

Construction Sector Deal 2017

Finally, we come to the financial deal that was struck between government and industry, effectively the culmination of all the research and industry-wide con-sultation that led to this point. The Industrial Strategy was the government's bid to boost productivity across the board by investing in '*Ideas, People, Infrastructure, the Business Environment, and Places*' (HM Government, 2017a, p. 14). As part of that programme, the Construction Sector Deal has delivered £170m of investment from the Industrial Strategy Challenge Fund (ISCF). This has been matched against £250m from industry to further digital tech-nologies, including BIM; new manufacturing technologies and production systems; energy generation and storage technologies; and R&D and demon-stration programmes, all with the education, housing and infrastructure sectors being specifically mentioned as key targets for these initiatives.

The Farmer Review is cited in this report as highlighting a combination of factors behind the industry's poor productivity levels, calculated to have fallen 21% below that of the wider economy since 1997. The reasons for this are stated as including '*the cyclical nature of the sector, the unpredictability of*

Lower costs

33%

reduction in the initial cost of construction
and the whole life cost of built assets

Faster delivery

50%

reduction in the overall time, from inception to
completion, for newbuild and refurbished assets

Lower emissions

50%

reduction in greenhouse gas emissions
in the built environment

Improvement in exports

50%

reduction in the trade gap between total exports and
total imports for construction products and materials

Figure 2.3 Construction 2025 targets.

Source: HM Government, 2013, p. 5.

future work and a lack of collaboration across the sector', with transforming the industry requiring '*shared leadership by the industry, its clients and the government*' (HM Government, 2017b, p. 6).

The Construction Sector Deal builds off the Construction 2025 targets, first published in 2013:

In addition to this there are the four 'grand challenges' that it is suggested we need to be focused on, all of which can help to deliver on these targets, and they are defined as AI and Data Economy; Clean Growth, the Future of Mobility; and our Ageing Society.

These are all well thought through and considered objectives, but we are now well over half way through the time allocated for them to be met, with no sign of any measurable progress having yet been made. This is not to say that targets are not needed, but that without the necessary infrastructure in place to know how they might realistically be achieved or even measured, they are, at best, meaningless, but even potentially damaging in the erosion of faith that stems from them continually being found to be unachievable.

There are, nevertheless, projects underway that aim to deliver on some of the objectives categorised under these four headings, some of which will be discussed in detail in Part II, but the three main themes focus on

1) the further development of digital technologies;
2) the promotion of off-site manufacturing (OSM); and
3) a move towards whole-life asset performance.

These are three well-targeted and interrelated objectives that together will undoubtably improve the productivity of the industry – if they are adopted. They would also help to resolve the three reasons given by Farmer for the currently low productivity levels:

1) the lack of industry collaboration;
2) the unpredictability of future work; and
3) the cyclical nature of the sector,

but only once the connections have been made, and the consequences linked back to the root causes.

To take those three themes and suggested reasons for low productivity in turn:

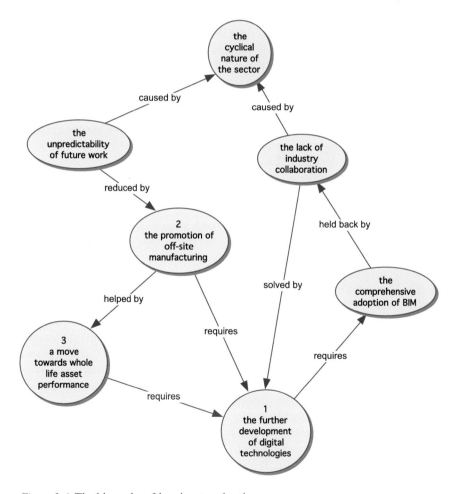

Figure 2.4 The hierarchy of barriers to adoption.

1) The adoption of digital technologies starts with the comprehensive adoption of BIM, a prerequisite for any true collaboration between the industry providers and their supply chains;
2) the unpredictability of future work, in part due to the cyclical nature of construction within our economy, could be greatly reduced by broadening the delivery model through a greater emphasis on OSM; and
3) OSM itself could be better promoted through its ability to deliver built solutions that capture the broader societal and environmental benefits that a whole-life costing methodology would measure.

Within these connections lies the untapped potential of how change can be brought about, not through coercion and persuasion, but through a clearer understanding of how that change can be of genuine benefit to those being asked to adopt it.

A deeper understanding of these and many more connections is therefore critical to these policies being adopted, but even that is only step one. Before any new innovation – product or process – can be realised as a mainstream solution, there are three barriers that need to be overcome – knowledge, motivation and ability, which are discussed in the next chapter. These are stages that must be met not just at an industry level but at a business level, and even at an individual level within that business. Changing the mindset of an industry starts with changing individual minds, and that is the level of implementation that is missing, along with the knowledge of what motivates those individual minds to make the decisions they do.

Conclusion – the Drivers for Change

There is an unintended bias to government initiatives that is limiting their impact. This begins with their initial remit, flows through the scope of the research undertaken, before finally being reflected in the recommendations made on completion. In addition to this, and to some extent as a consequence of this bias, minimal attention is paid to how these policies are then to be implemented 'at the coalface', beyond them being handed over to the industry or its representatives to deal with. The single-point perspective solutions that are generated out of this mindset have repeatedly failed to capture the imagination or support of the stakeholders that need to engage with them to ensure any mainstream adoption, resulting in the government resorting to coercive methods – carrots and sticks – and the perpetuation of the industry's familiar 'us and them' relationship with government.

In the next chapter we will see just how many different perspectives there are that are not being considered, and how they can explain the industry's lack of engagement, so often put down to complacency, but in reality, more an inability to participate. If there is complacency in the process, it could in fact more appropriately be levelled at successive governments for not attempting to fully understand the industry they are so keen to transform.

Bibliography

BLP Insurance. (2018). *Progress made but more urgency required: The construction industry 18 months on from the Farmer Review. Presented by guest speaker Mark Farmer.*

Cant, C. (2019). Problems with service of CIL documents. *Local Government Lawyer*, April. www.localgovernmentlawyer.co.uk/planning/318-planning-features/40291-problems-with-service-of-cil-documents

Construction Leadership Council (CLC). (2021). *CLC Strategy 2021.*

Department for Communities and Local Government (DCLG). (2017). *Fixing our broken housing market*, February.

Designing Buildings. (2023). *Construction industry reports.* www.designingbuildings.co.uk/wiki/Construction_industry_reports

Farmer, M. (2016). *The Farmer Review of the UK Construction Labour Model.* Construction Leadership Council. www.constructionleadershipcouncil.co.uk/news/farmerreport/

Grayston, R. (2017). *Slipping through the loophole: How viability assessments are reducing affordable housing supply in England.* Shelter, November. https://assets.ctfassets.net/6sxvmndnpn0s/4CzQDIXrSQXLFGlIGZeSVl/634340e83ce2cb1e2dec9879d19a5db3/2017.11.01_Slipping_through_the_loophole.pdf

HM Government. (2013). Construction 2025. Industrial strategy: Government and industry in partnership. *UK Government*, July, *78*. https://doi.org/HMGovernment

HM Government. (2017a). *Industrial strategy: Building a Britain fit for the future.* November. https://doi.org/10.1049/ir:19930092

HM Government. (2017b). Industrial strategy. Construction Sector Deal. *IEE Review*, *39*, 5. https://doi.org/10.1049/ir:19930092

Letwin, R. H. S. L. (2018). *Independent Review of Build Out.* https://assets.publishing.service.gov.uk/government/uploads/system/uploads/attachment_data/file/752124/Letwin_review_web_version.pdf

Madrian, B. C. (2014). Applying insights from behavioral economics to policy design. *Annual Review of Economics*, *6*(1), 663–688.

Martin, J. (2021). *Productivity in the construction industry*, UK. Census 2021, Office for National Statistics.

Mitchell, A. (2022). Leadership council spreads its wings and morphs into something else. *The Construction Index*, November. www.theconstructionindex.co.uk/news/view/leadership-council-spreads-its-wings-and-morphs-into-something-else

Moore, G., & McKenna, R. (1999). *Crossing the chasm.* HarperCollins.

3   Barriers to Change

The Industry Perspective on What the Government Needs to Understand

For change to take place, there needs to be knowledge, motivation and ability: Knowledge of the alternative solution being proposed, motivation to engage with it as an idea to be considered, and the ability to make that transition should the motivation be great enough. Just these three hurdles to clear when promoting an innovative solution. Dig a little deeper, however, and the sheer enormity of the task of challenging the status quo soon begins to look overwhelming. Talking down the chances of success might seem a strange way to drive change, but it is only through understanding where the barriers are set that they can be dealt with – through removal or avoidance – and a route forwards found, represented by the path of least resistance. Too often the construction industry has been set up to fail by being given a challenge where only some of the exogenous barriers have been considered to be relevant, let alone removed, to pave the way for a successful outcome. If the end of the road remains blocked by a barrier that had been deemed either too difficult to deal with or peripheral to the problem as defined from the initiator's 'single-point perspective', the end result will always be the same.

Knowledge, motivation and ability

The construction industry has been defined as both adversarial and fragmented. It has over the past 40 years developed a 'survivalist shape' designed to help it deal with those external political and economic realities over which it has little control. The trajectory on which the industry has been set for so long now has resulted in siloed thinking, under-informed decision making and, as a consequence, low productivity levels. Predominantly small SME businesses competing within long supply chains define the bulk of the industry and it is not an environment conducive to collaborative working relationships. Neither is the cyclical nature of our economy, felt more strongly and immediately in construction than in most other sectors, conducive to long-term employment prospects or investment in R&D.

DOI: 10.1201/9781003332930-5

These are structural issues that cannot be resolved from within, and if left unresolved, no amount of political coercion is going to have the impact needed to change how the industry operates within their constraints. This is an industry made up of over a million small businesses (Turner, 2019). They might have the knowledge, and in some cases they might have the motivation, but in many cases they just haven't got the ability to change course, even once those first two hurdles have been cleared. This is not to say, however, that change is outside the reach of these businesses. What is outside their reach is the kind of change envisaged by those enacting change agendas from above, with little understanding of what it is that these businesses need to get from adopting change. There are still many unknowns, and the next task is to define some of the as yet unrecognised barriers and explain why it is that initiatives that fail to define those barriers go on to fail in their implementation, by constantly running in to them.

Neither is this saying that the motivations driving these initiatives from a governmental perspective are necessarily wrong – we do need to increase the industry's productivity levels, and we do need to change the way we build to do that, and also to allow new sustainable solutions to take root. What is so often wrong, however, is the approach taken to enact or deliver these initiatives, because in most cases it will be based entirely on the government's objectives. Sometimes these initiatives can stray irreparably off target, but more often it is little more than semantics standing in the way of successfully translating a lost message into an appealing proposition – the difference between talking 'productivity' or 'profitability', or between a 'rulebook' and a 'guidebook'. It might seem like a long and onerous task, but the education of all parties invested in our construction industry, including the government, about each other's needs and motivations, is most likely the short cut to successful change.

Exposing the Barriers to Change

One of the agendas outlined in the Construction Sector Deal was the development of a common metrics methodology to help gather comparable data on the benefits of Modern Methods of Construction (MMC). What seemed at the outset to be a straightforward and uncontentious task, however, has yet to be resolved, and is discussed in detail in Part II. One of the many issues that came to light, and the one relevant here, was that the metrics chosen for this task were only ever those that showed the *benefits* of adopting MMC solutions. For a business considering such a radical change to its working practice, the potential benefits alone are not enough. Far more important are the costs that will be incurred in reskilling or the purchase of new plant, the level of risk involved in new processes and customer perception, the unknown reliability of the new supply chains, the possible time delays due to accreditation processes – all these factors are real and measurable but are never openly discussed.

This is another example of the single-point perspective approach to problem solving at work. If the problem is couched as one of 'how to convince the industry that MMC is a better solution than traditional build', it is not surprising that only the positives will be brought into play. But this is from the perspective of the government's desire to boost productivity, broaden the delivery base, and to some extent help meet health and safety and sustainability targets. These are not the key drivers of small companies trying to sustain their profitability from one month to the next. Understanding their drivers might help explain why MMC solutions are seen as less attractive than expected, and more importantly show how the product, or maybe even just the message, might need to change to address this.

If the benefits of adoption are a measure of a solution's *viability* when compared to the status quo, the barriers to adoption could be seen as a measure of that solution's *feasibility*, the practical considerations more closely related to the real-life consequences of making that decision. These might not be so immediately obvious to the outsider looking for a wholesale transformation of the industry to boost productivity, but they are as important if not more so to the businesses being asked to make those changes.

Returning to the three barriers to be overcome, knowledge, motivation and ability, to really begin to understand the rarely discussed concerns they encapsulate, they can each be further subdivided to reveal even more levels of choice that exist within this decision-making process.

Knowledge:	**Awareness**	Motivation:	Ability:
	Authenticity		
	Subjectivity		

Awareness: With near universal access to the internet, knowledge would seem to be a freely available resource, and a negligible barrier to overcome, but without an awareness of a product's existence there will be no knowledge. In a crowded space and with limited time for research, finding something, even when that something is trying to be seen, can introduce an early limiter on the pool of products being chosen from. Creating that awareness is therefore a major hurdle, and a cost to be factored in to the promotion of any new product, process or initiative.

Authenticity: Not all data sources are reliable, and the easiest to find are often the least likely to be accurate or thoroughly researched. Tracing information back to its source for verification is a time-consuming task and a luxury rarely available to industry decision makers. Search engines are very efficient at directing users to the information that providers want them to see, and then displaying that information in the most accessible format to minimise the time needed to arrive at a decision. Misinformation is therefore a very real threat to both the promoters and the adopters of new innovative solutions.

Subjectivity: The selective use of data to only focus on the positives not only conceals the downsides, but makes any comparative assessment very difficult to measure, since no two alternatives are necessarily going to use the same benchmarks. This skewing of the data is further compounded by users rarely looking beyond the first option, which will inevitably appear to answer all the questions asked of it – by the company with a vested interest in promoting it.

Even independent sites looking to provide balanced advice can struggle to find the shared denominators that will allow for a fair comparison. And in addition to this, the power of perception over factual information cannot be overestimated. What is perceived to be an issue or a problem can live far longer in the public consciousness than the reality of improved solutions.

Figure 3.1 gives an insight into how the housing industry sources its information, and how that can vary depending on the information being sought.

Information Sources Questionnaire

As part of my research I need to establish where the housing industry gets its information from. The following questions are set out in a format that will allow for direct comparisons, but will not necessarily work for everyone, so please ignore boxes that are not relevent to your sector.
0 = not applicable   1 = slight   2 = average   3 = considerable   - or just a tick or cross if that is more appropriate
please state your industry role:

SUBJECT MATTER:

What subject areas are of most concern to you?	materials / method of construction	policy / legislation	house building / company news	public opinion / trends	comments / other
			12	14	

SOURCE:

Where do you usually find this information?	Academic papers	Government publications	Industry literature	general media	comments / other
materials / method of construction	8	11	13	6	
policy / legislation	2		12	11	
house building / company news	2	5	18	13	
public opinion / trends	2	7	9		

INTERPRETATION:

At what level do you get this information?	full publication	executive review / summary	commentary / analysis	in conversation	comments / other
academic papers	1	12	6	4	
government publications	18	14	10	5	
industry literature	13	11	9	4	
general media	4	5	7	6	

FORMAT:

In what format do you access this information?	printed form	websites	digital media	conferences / meetings	comments / other
academic papers	0	13	9	6	
government publications	0	16	9	10	
industry literature	6	15	9	10	
general media	2	12	14	1	

VALIDITY:

How much of your knowledge do you feel is down to:	cumulative experience?	original thought / innovation?	trust in authorities / procedures?	gut instinct / perception?	comments / other
materials / method of construction		8	8	5	
policy / legislation	7	1	14	1	
house building / company news	6	1	9	3	
public opinion / trends	6	1	9	10	

Figure 3.1 How we access information.

The limited feedback received in this survey means this is very much a tool to be used rather than a set of results to be acted upon, but even this limited response begins to show how important such an exercise can be in gathering empirical evidence to either confirm or refute the perceptions thought to exist. These are a few key examples of the insights gained.

Areas of interest: Broad, with a higher than expected interest in public opinion but a reliance here on general media as a source together with gut instinct and perception.

Government policy: There is a high level of trust in information accessed on government policy, which is most often read at source, rather than through opinion pieces, but always digitally or in a conference setting.

Academic research: Academic papers are rarely read, but when they are they tend to be on materials and construction, executive summaries only, and usually digitally.

Communication: Printed literature is rarely used compared with the internet as an information source, and discussion between individuals and original thought play a minimal role in the decision-making process.

This throws up some interesting questions. Why is it that industry information, when viewed through the lens of government reports, is trusted far more than the original sources used to inform those government reports in the first place? How can complex arguments win over simplistic solutions when most decisions are taken on the basis of executive summaries only? Who is providing the public's opinion beyond a general perception of what they are thinking?

Trust, simplification and perception are heavily relied upon methods for arriving at decisions, and perversely, the more complex the problem, the more this has been proven to be true. The research on this proved that the time taken to make decisions is based on the inelastic attention span of the decision maker, resulting in complex decisions being inevitably more poorly considered (Toma & Gheorghe, 1992).

'Equilibrium and Disorder in Human Decision-Making Processes'
(Toma & Gheorghe, 1992)

Decisions often begin as an internal exercise based on individual research, even if the incentive to do so comes from some external event. In this paper, the authors demonstrate that there is an optimum decision making time based on information availability, but that in reality the time taken to make decisions is based on the inelastic attention span of the decision maker, resulting in complex decisions being poorly considered. The paper suggests that too much information becomes counter-productive as the decision maker, unable to select the really important issues, becomes unable to reach a satisfactory conclusion within the time available, resulting in either inappropriate decisions or a reluctance to take any action at all.

'In short, there are many reasons to expect that, on their own, individuals (either lay or expert) will often experience difficulty making informed, thoughtful choices in a complex decision-making environment involving value trade offs and uncertainty' (McDaniels et al. 1999).

This is certainly an apparent phenomenon in the house building industry where the rate at which regulations change and building solutions are accredited to meet those regulations means that the evolutionary process of technologies being honed from experience over time has been replaced by a system of guarantees and indemnity cover. As suggested in this paper, the time available to decide upon, let alone trial a solution is seldom adequate, and the level of research carried out reflects this. 'Information multiplies our connections and forges greater interdependence' resulting in a pressing need 'to develop better monitoring systems and faster reaction capability' and to 'develop a critical attitude toward disinformation and information manipulation'. Under our current system of operation, 'we are not able to accurately process the information we gather and to reach adequate conclusions because we gather these data from an informational "environment" that seems familiar but basically is "incomplete"'.

There is no solution offered in this paper in response to these observations, but it is something that this research needs to address in a realistic manner. If the assumption is made that the housing industry will continue to be a complex environment and the time available for decisions will continue to be limited, it is only the decision making process that can be improved to meet these challenges.

(Siebert, 2019)

Knowledge:	**Motivation:**	**Economy**	Ability:
		Equity	
		Ecology	

In this instance, what motivates a business to make a decision has been broken down into the three well-established categories known as the 'three dynamics of sustainable communities' – economy, equity and ecology (Hansmann et al., 2012); financial benefit (short term), societal benefit (medium term) and environmental benefit (long term). Immediately, it becomes apparent where the natural instinct lies for a business focused on its own profitability, where immediacy will invariably trump longer-term ethical considerations.

Economy: The simple way to get new innovative solutions adopted is to make them less expensive than the existing solution, something that is increasingly difficult to achieve. Many improvements to current ways of working invoke some degree of broader societal or environmental benefit, which will inevitably 'pay back' over longer timeframes than that of a single project. This payback period is therefore a critical element of the information needed to be able to calculate viability. But even a cost-neutral alternative requiring no such calculations incurs upfront costs in terms of retraining / reskilling and the possible purchasing of new tools and equipment, all of which also require time and resources to implement.

Equity: The challenge of operating in a fair and equitable way was once seen as a sure route to unprofitability, but through a combination of altruism, philanthropy and the unionisation of the workforce, societal improvements were eventually understood to 'pay dividends' through a combination of better working conditions, higher employee satisfaction and the social standing of businesses achieved amongst the public at large due to these commitments. The degree to which those dividends return directly to the business deciding to act equitably, however, varies with the measures taken. The impact of employee incentive schemes or improved working conditions can be directly measured over time, if not immediately, whereas, for instance, reducing the inconvenience caused to the general public during the construction of a building can be harder to translate into bottom-line profit in any way other than a negative cost outcome. It is in these areas that the indirect benefits accrued need to be defined in ways that are measurable and relevant to an individual business's decision-making process.

Ecology: If financialising better working conditions as a benefit to a business's profit margins was seen as a hard sell for the previous century's trade unions, making the same case today for introducing environmental measures becomes far harder, with the benefits being defined in generational terms rather than just a few years. It also becomes harder to realise the benefits at an individual business level, creating an even stronger need for collaboration and enforcement, but even more than this, a need for education. The same rule applies, however, in that the benefit in making an environmentally motivated decision, like an equitable one, still needs to be defined in terms of direct financial benefit before it will be seen as justifiable by the vast majority of businesses. This is pragmatic idealism – not losing sight of the ultimate goal, but equally not denying the reality of what most businesses have to prioritise – their own immediate profitability.

Knowledge:	Motivation:	**Ability:**	**Disunity**
			Complexity
			Uncertainty

Perhaps the least understood barrier to change is the *ability* to enact something, even after its benefit has been recognised and proven. Ultimately this is due to the risk that change entails, and how that risk is shared between government and industry, between the industry stakeholders who need to act together to ensure a successful transition, and even between the individuals within a business, one of whom will have to decide to sanction the decision to 'jump'.

Disunity (internal): The ambiguous risks of internal divisions; The fixed time available to make decisions

This first category of risk refers to internal risks, or risks than can at least be resolved without the co-operation of others outside the organisation – but can be as difficult and divisive as any other. They tend to involve the dissemination of information to ensure that everyone is on the same page, and understands the reasons behind the decision being taken. This also ensures that the full implications of what that change means at every level have been understood and allowed for by those making that decision. (See the example in Chapter 4 regarding the attempted introduction by one firm of Large Format Blocks.)

Complexity (industry level): The barriers to implementation; Hierarchy of decisions; Conflicting motivations; Supply chain complexities

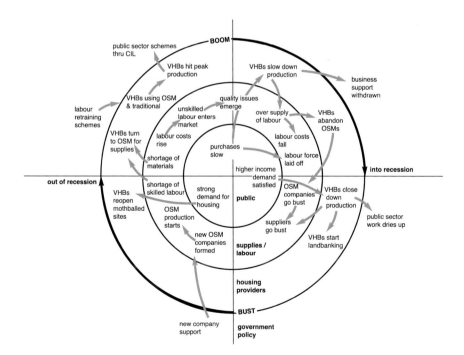

Figure 3.2 The cyclical adoption of off-site manufacturing.

Because of the complex supply chains that define the construction industry, there is also a level of risk associated with getting all stakeholders to commit to adopting a new way of working simultaneously. The torturously slow implementation of BIM in the UK is a case in point, where the payback from investing in BIM can only be fully realised once the whole supply chain has committed. Conflicting motivations can also destabilise the transition from one system to the next, as in the adoption of off-site manufacturing over traditional build methods by the housing sector. The necessary and ongoing commitment this requires from the volume housebuilders has never materialised beyond a brief and recurring interest at the end of recessionary cycles when there is a peak in demand for housing and a shortage of on-site skilled tradesmen.

There is much to unpick within this cycle of partial adoption that will become clearer as we get into resolving the issues that underlie this pattern of behaviour, but the most important message to take at this point is the sheer number of barriers that off-site manufacturing will have to overcome *simultaneously* before any mainstream transition can be expected to materialise.

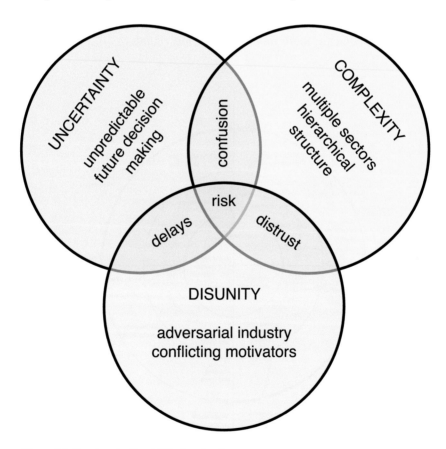

Figure 3.3 Barriers to the ability to adopt.

Uncertainty (external influences): The disconnect between risk and reward; The role of uncertainty – political and environmental; Risk associated with a cyclical economy

This third tier of risk exists at a level that transcends the construction industry itself and is therefore the hardest to manage. Political and economic conditions are key to the successful promotion and adoption of an innovation, and although difficult to influence, cannot be ignored. Changing these macro-economic conditions under which the industry operates is not an impossibility, but it does require a whole new level of collaboration and political interaction that requires time and a level of dedication that needs to be fully understood at the outset. The alternative solution is to accept these exogenous barriers as 'immovable' and factor them into the equation as such. In this way, even if these factors are negatives, they are no longer uncertainties as they have now been allowed for, thereby reducing the level of risk they contain. It is also worth noting that there is a difference between uncertainty and volatility at an economic level. The cyclical nature of our economy is not in itself a detriment to innovation if that cycle can be predicted and allowed for (Roper & Tapinos, 2016). Taking a calculated risk is very different to taking a chance.

Applying the 'Barriers to Change' Approach

Within these categories lie all the reasons why change is so hard to bring about, why well-meaning initiatives fail to live up to their promise and why the status quo can retain its dominance well beyond its apparent sell-by date. But it's not actually these barriers that are preventing any progress so much as the failure to address them as barriers, and to instead focus attention solely on the benefits to be gained from the changes being promoted. The aim here is to show how, by turning our attention to the reasons why change is not as well received as we feel it ought to be, we can bring about that change far more effectively than we have been able to by only focusing on and promoting its benefits.

Before applying this approach to specific industry problems, however, we can first see how much more effectively the Farmer Review's findings could have been acted upon had the government's and the industry's interpretive net been flung a little wider, and some of these supposedly peripheral concerns brought into the discussion.

At the outset it was suggested that the Farmer Review might have reached some more fundamental but potentially 'difficult to act upon' conclusions if its remit had been wider, and similar comments were then made about the subsequent papers and analyses published in response to it. In reality, all research has to be limited to some extent, whether that is at an individual level within a company choosing what product to specify, or at a governmental level, choosing between policies to enact. But at any level, the consequences of decisions based on incomplete evidence are the same, and far more expensive in the long run in terms of time and resources than any increase in expenditure at the research stage. How to define the scope of that research at the outset to ensure all relevant factors are captured is therefore critical to the conclusions that will be reached.

Define the scope of the question being asked

The Farmer Review purported to have sought the root causes behind the 'symptoms of a dying patient' but claimed one of these to be the industry's adoption of its 'survivalist shape'. There is a very simple method for assessing whether a root cause has been reached or whether the search for that cause is still working its way through a hierarchy of consequences, and that is to simply ask 'why?' In addition to 'drilling down' to establish the root causes to a problem, it is important to define the period of time over which that problem has developed, giving us both depth and length as constraints to our research. And the final way in which the scope of the questions needing to be asked must be defined is: Who is being asked those questions? How many different groups of people or stakeholders do we need to put the same question to before we stop getting different answers? That now provides the third dimension of breadth and a clear set of parameters to work within, and also with which to assess the validity of any research carried out and applied on the industry's behalf.

Research can be defined by its depth, length and breadth

The Farmer Review carried out by CAST, an independent consultancy, was free to set its own parameters and clearly stated its intention to build off past experience: '*Critically, a plan for change needs to recognise, based on past evidence, that the industry will not change itself unilaterally at scale*' (Farmer, 2016, p. 10). Farmer also chose to look outside the usual constraints of the industry for solutions: '*the most important and effective drivers for change do not necessarily sit within the industry itself*' (Farmer, 2016, p. 4). That gave the review both length and breadth, but it is the depths to which this research was prepared to go to reach its conclusions that possibly limits its ability to bring about the level of change anticipated, and it is the reasons for that limitation in scope that need to be questioned.

The approach taken here to 'dig a bit deeper' has been to try and put these symptoms into a hierarchy that can begin to suggest what those more fundamental, underlying problems are. Farmer suggests that there are ten symptoms in total to be addressed and that they stem from three root causes, which he defines as a lack of a strategic framework, a non-alignment of interests across industry and the industry adopting a 'survivalist shape'. All three of his root causes sit within the confines of the industry itself, and whilst the government is not excluded from the task of fixing the consequential problems defined in Figure 3.4, its actions are excluded from being considered as creating those problems in the first place.

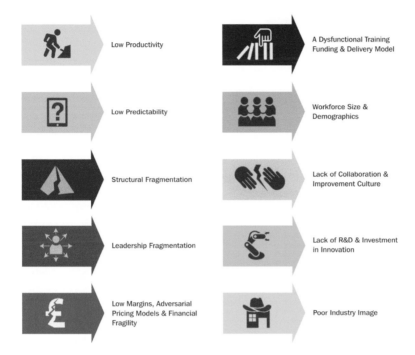

Figure 3.4 The Farmer Review's ten symptoms.

This is not a blame game, but just an attempt to capture all the factors that need to be included to fully inform the questions being asked before we can arrive at any meaningful answers. The answer may well be that government policy 'is what it is' and sits outside our sphere of influence, but accepting a barrier to change is very different to ignoring its existence or its ability to impact on the choices that can be made. Fundamental to this is the realisation that barriers can be either removed, avoided or accepted and all are equally important choices in the decision-making process.

In this instance, there are clearly some systemic economic problems that have led to the industry adopting a survivalist shape, and these are well known. Our cyclical economy has repeatedly taught the construction industry, and the housing sector in particular, not to rely on a regular supply of work, or more precisely, on clients able to sustain the demand needed to work profitably. This is the root cause behind the housing industry's chosen business model, where employees are seen as dispensable, long-term training is seen as unprofitable and a transition to factory-based production, with all the long-term overheads that implies, seen as unsustainable.

This has already taken us squarely into the realms of political decision making and outside the normal boundaries of acceptable discourse, certainly for a government-commissioned report on how to solve problems at an industry

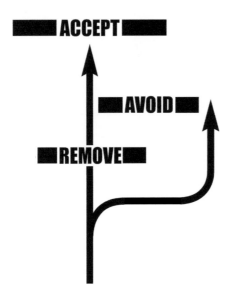

Figure 3.5 The options ahead.

level. But where does that get us in reality? And have we even reached the root cause, or is there still more explanation needing to be uncovered? Why, for instance, do we have such a cyclical economy? Is it normal and is there anything we can realistically do about it? Getting used to every question leading to more and more questions is part of this process and quickly reveals how often we fail to question anything to its full extent. As a child, the game of 'why' would invariably end with a parental 'because I say so', but we appear to have grown up accepting that at some point this will always be the answer we should accept.

Most questions lead to more questions, not answers

The reality is that not all countries have equally cyclical economies, because some countries' governments are more interventionist and do more to balance out supply and demand between boom and bust cycles (Iacono, 2018). In the case of housing, this is exhibited in a tendency for some countries to build more social housing during recessionary periods when labour is cheaper and the demand for market housing is less. This in turn requires governments to limit spending during a boom time to below what would be possible in order to build up the reserves necessary to fund that social housing programme. To ask the question 'why do we have such exaggerated boom and bust cycles?' would therefore result in an even more politicised answer relating to short termism and political ideology / expedience.

Revisiting Farmer's Ten Symptoms

To conclude, where does that get us that is of any more use in terms of solving the problems at an industry level? The diagram in Figure 3.6 incorporates these suggestions and shows how the ten symptoms defined by Farmer sit within a

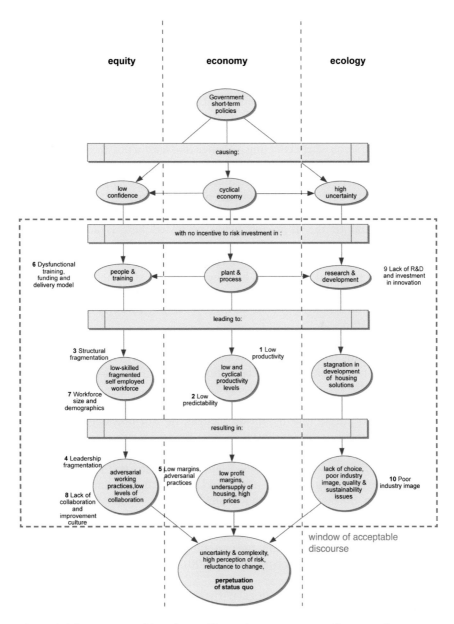

Figure 3.6 Root causes of inaction vs. Farmer's ten symptoms of poor performance.

hierarchy of causes and consequences, and could be shown not only to stem from one fundamental root cause, but also to all result in the same manifestation, that of inaction:

Whilst the Farmer Review sets out how the industry must deal with these ten symptoms, the proposition made here is that the root causes of these symptoms are the critical factor, and that they must either be addressed or accepted, but they cannot be ignored. Eighty years of white papers failing to fully resolve the issues they set out to address would suggest that not only are the root causes a constant in these equations, but that they will continue to be so, and should possibly be accepted as factors 'outside the bubble' of what can readily be influenced. Re-evaluating the Farmer Review on this basis leads to some interesting conclusions, as there is an element of this pessimism / realism to be found in many of the review's findings, but not expressly dealt with as political problems to be confronted. Without saying as much, Farmer chooses to focus on the areas where effecting a change is at least a possibility, and treats those that are outside the industry's sphere of influence as 'unchangeable' constants, no matter how undesirable that may be. The problem with adopting this approach, however, is that the impact of those exogenous factors is effectively being ignored rather than accepted.

Fundamental to the approach being proposed here is accepting that our form of government results in a number of predictable outcomes that would require systemic changes to our political system to influence. This is not a political discourse, but the following statements need to be acknowledged because in principle they are accepted realities and also areas where there is little likelihood of change occurring through any industry-scale intervention:

- Our democratic five-year cycle encourages short-term policies that to some extent accentuate the economic conditions that lead to our boom and bust economy (HM Government, 2015, p. 512).
- The reality of the public's voting habits means that collective benefits, whilst easy to incorporate into manifesto pledges, do not necessarily guarantee electoral success.
- Bold, easy to understand policy directions, such as New Towns, off-site manufacturing, the Localism Bill, Right to Buy, etc., will tend to be used to placate the voting public so as to avoid confronting the underlying issues around our shortage of housing (Mann, 2017).
- The need for governments to be impartial when promoting a specific option within any sector of the economy results in an inability to dictate the direction of travel for any meaningful period of time (interview with G. Thompson, 2016, in Siebert, 2019, p. 120).

In addition to this, specific political parties can be expected to behave in their own ideologically driven ways, and this too needs to be taken into account.

An acceptance of these political barriers exposes the limitations of the Farmer Review's findings. By way of example, the volume housebuilders are businesses and have always behaved in a predictable manner. Their terms of reference are very clear and do not involve solving the nation's housing crisis through the adoption of new innovative solutions that, whilst they might increase the quality of their product or their output potential, will not necessarily grow their profit margins. To this day, governments still seem to find the speculative housebuilders' business model difficult to come to terms with (Baseley, 2016). Again, the Farmer Review begins to recognise these truths and there is evidence in its recommendations of how this particular issue should be worked around rather than tackled head on:

> *More tenure diversity would immediately imply different supply chain and delivery models that may better promote innovation. In time, this may in turn influence core housebuilder delivery models but it is considered unlikely that large scale innovation will start in the volume housebuilder market.*
>
> (Farmer, 2016, p. 9)

This is an 'accept and avoid' approach. It accepts the volume housebuilders' business model and avoids dealing with it by suggesting 'more tenure diversity' – i.e. more rental property provision – as the solution to promoting innovation, which would be a legitimate strategy if it were extended to define how that additional rented housing stock was to be delivered. Rental housing in the UK comes in three flavours – social housing (set at around 50% of market rent value), provided primarily by local authorities, 'affordable rent housing' (no more than 80% of market rent value) provided by housing associations, and market rent housing provided by the private sector – although increasingly local authorities and housing associations are having to build some of their housing for market rent to subsidise their affordable housing provision. The unresolved problem lies in the fact that none of these models act independently of the volume housebuilders' model, with the vast majority of social and affordable housing provision linked directly to it, making any counter-cyclical delivery to help ameliorate the peaks and troughs of the economy hard to achieve. Private rental provision, on the other hand, works to a very similar business model to that of the volume housebuilders, with bottom-line profit being a far stronger motivating factor than any societally or environmentally driven agenda.

The reality of this solution, therefore, is that, whilst it attempts to avoid the barrier of the volume housebuilders' business model, the alternative route taken still runs into the same issues born out of that sector's continuing market dominance. To genuinely achieve more tenure diversity requires strategies that step outside the boundaries of acceptable discourse and consider the political decisions being taken at a broader macro-economic level. Crucially, this does not mean that the solutions being sought need to be taken at that political level, but for any decision to be effective, it has to be based on a full understanding of the political environment in which the industry has to operate.

A quick look at all ten of these symptoms shows how each can be explained by factors, mostly political, that do not feature in the Farmer Review but which radically change the conclusions reached about how to address them:

---

1 Low Productivity

Productivity has increased substantially in all other industries other than agriculture in the past 50 years, whereas in construction it has fallen. Previous reports have reprimanded the industry for this and suggested that it should behave more like other industries, by adopting more automated processes (Egan, 1998, p. 11). A more profitable approach might be to look at how the construction industry differs from these other examples and to understand whether or not it can be automated to the same extent. The labour element is clearly a component in the low productivity measures, as it is with agricultural industries, but the cyclical economy also impacts on this more so than in other industries, with unskilled labour having to be brought on line

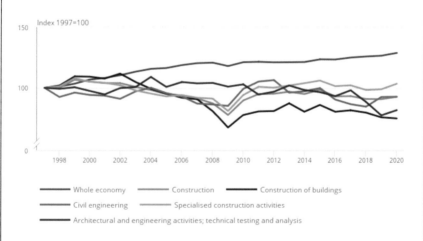

Output per hour worked, construction industry and sub-industries and whole economy, UK, 1997 to 2020, index 1997 = 100

Source: Office for National Statistics – Labour productivity

Figure 3.7 The construction industry's productivity deficit.

---

during boom periods, with a consequential drop in productivity and quality (Farmer, 2016, p. 14). A move to off-site manufacturing, the standardisation of components and also of the finished product would all move the housing industry in the right direction, but the root cause of many of the issues that have led to the housing industry's current low productivity levels, i.e. the cyclical economy and the exaggerated impact that has on housebuilding, will still beset the industry and potentially prevent a high-investment off-site manufacturing sector becoming fully established, as it has in the past.

[Low productivity tied to the cyclical economy]

2 Poor Predictability

Two fundamental differences between the housing industry and the manufacturing industries with which it is so often compared are the unpredictabilities of both the sites being built on and the weather. These variables make the predictability of manufacturing and delivering houses on time, especially by conventional means, a very different proposition to any factory-based manufacturing process. But statistics for low-rise against high-rise construction show that another factor might also be at play, as the successful on-time completion rates for low rise, much of which is housing, are far higher than for high rise (Farmer, 2016, p. 16). Poor quality of work, budget overspends and missed completion dates are more often than not put down to skills shortages, the effects of which are more pronounced in complex constructions than in conventional housing, where temporary low-skilled labour can be brought in at short notice when necessary. So the root cause, it is being suggested, beyond the unpredictabilities of site conditions and the weather, is the cyclical economy and the associated problems of long-term investment in the training of its skilled workforce. If the hope is that off-site manufacturing will reduce the problems associated with poor predictability, it will also need to address the underlying issue of cyclical demand by compensating for this during downturns.

[Poor predictability is tied to the cyclical economy]

3 Structural Fragmentation

The construction industry is known for its lack of collaboration and complex transactional processes. To some extent this is due to the fact that construction is a complex multi-disciplinary process, but it is exacerbated by the high levels of self-employment that add to the introverted nature of its component parts and poor levels of communication between them. This long tail of SMEs making up a large, if declining, proportion of the housing industry (Communities and Local

Government Committee, 2017, p. 13) has resulted in the UK housing market being hard to penetrate by overseas markets, keeping innovative European methods of construction at bay (Farmer, 2016, p. 17). A culture of sub-contracting has also increased the complexity of the supply chain and resulted in layers of process management that add to overall costs and inefficiencies. This is a good example of where the 'survivalist shape' of the industry can be seen to emerge, and also where the Farmer Review suggests that there are alternatives to correcting these now endemic practices, but again, the alternatives are fixes rather than true long-term solutions: Laing O'Rourke is used as an example of the 'New Entrants' who, rather than trying to improve the existing 'accepted business model', are taking this alternative course of action, where they can avoid this fragmentation by setting up and controlling their own supply chains, and by establishing their own vertically integrated markets around off-site manufacturing. This effectively reduces the company's risk by removing the uncontrollable contractual elements of the building process. What this doesn't remove, however, is the risk associated with a cyclical economy, and by investing in an employed labour force, not to mention the considerable off-site manufacturing plant costs, that risk could, if anything increase. The issue of maintaining a regular throughput during a recessionary period therefore remains the unresolved factor in the equation.

[Fragmentation is endemic and also tied to the cyclical economy]

4 Leadership Fragmentation

The degree to which the industry is operating in pockets of self-interest is symbolised by the plethora of advisory bodies and lobby groups operating within the space between industry, the government and the public, often covering very similar ground. Add to these the umbrella groups that are attempting to gain more relevance by connecting these bodies in differing permutations, and the picture becomes one of an almost complicit fragmentation, with no real oversight to bring these efforts together. The Farmer Review recognises the need for these bodies to amalgamate under a common goal (Farmer, 2016, p. 20), but seems doubtful of the likelihood of this ever happening, as the industry is effectively in competition with its constituent parts. Again, there is reference to the modernisation that is needed coming from 'outside the traditional housebuilder approach', with the 'New Entrants' forging an alternative path being a more promising scenario than attempting to reform the industry from within. The government, on the other hand, whilst it is accepted that it does want to drive reform, is seen as ineffective, not only because of the limited influence it has over the wider construction industry, with the public sector works it controls put at only 25% of the industry's total output, but also because its own procurement system is as fragmented as the industry itself. Clearly the industry is struggling to co-ordinate its thinking due to the incredible complexity of its constitution, and to accuse it of lethargy, as this report does, is possibly missing a fundamental point. It is more likely that this lack of leadership and co-ordination has become the real barrier to change, not a lack of openness to change itself. If this is the case, the reason why the New Entrants are seen as more likely to embrace change might not be because they are less complacent, but because they have developed a simpler model that allows them to embrace change without the risk that comes from having to co-ordinate such a complex body of participants. The proposition being made here is that with complexity comes risk, and with risk comes an inertia to change, which is misconstrued by this and the many preceding reports as disinterest, lethargy and complacency.

[Fragmentation of control increases the risk associated with change to unviable levels]

5 Low Industry Margins, Adversarial Pricing Models and Financial Fragility

The review paints a picture of an adversarial business model with many non-value-added tiers within a supply chain that supplements its low profit margins by extending payment terms to enhance cash flow. Operating margins are low and cyclical along with the economy, which tends to promote opportunism when either material or labour supplies are limited. Competitive tendering has long been the norm within the industry, but this is not the first review to suggest that the industry needs to find a more collaborative way of working that encourages repeat business within the supply chain and promotes a more stable learning process, and as a consequence more profitable outcomes. Again, the sur-vivalist shape can be seen in the many ways the industry has learnt to 'play' the adversarial system rather than replace it, with '*whoever wins a project often being the party that has made the largest mistake in pricing it*' (Farmer, 2016, p. 24) becoming normal procedure. The choice, as suggested in the previous sections, is to either improve the system that exists or bypass the issues that are increasingly seen as systemic and too entrenched to affect. The Farmer Review itself accepts that:

> *The reality is that many clients, especially in the real estate development sector, are simply conditioned to operating in an adversarial way with industry and do not see a case to move to more collaborative and integrated approaches for fear that a lack of commercial tension will impact their own financial outcomes.*
>
> (Farmer, 2016, p. 24)

If this is the case, and to replace the long-standing system of competitive tendering is more a pipe dream than a realistic goal, the only alternative is to create the non-adversarial environment by taking as much of the operation as possible in-house and thereby control the production process from within. Again, this is implicitly described in the Farmer Review as the New Entrants' business model, without wanting to admit that the current housing industry's model is close to irreparable.

[The adversarial model cannot be changed by the industry alone and requires government to create a more collaborative environment within which to operate]

6 A Dysfunctional Training, Funding and Delivery Model

Training and skills development within the industry is funded by a levy raised by the CITB (Construction Industry Training Board) from all its member employers and redistributed through initiatives and grants. The housing industry profile is one with a long tail, where a large percentage of housing, and that using the most innovative methods of construction, is delivered by small SMEs who the review finds do not benefit proportionally from the grant system. To some extent this is due to small SMEs not having the resources to apply for the funding in the first place, but there is also a feeling amongst these companies that training is not cost effective because of the tendency of employees to move on once qualified, or drop out altogether during recessions. Overall, the level of training offered by construction industry employers is the second lowest of all industry sectors (Farmer, 2016, p. 25), and this feeds into the sense of low confidence and poor industry image amongst the general public.

Systemic changes are needed to rectify these problems, and they have been called for on many occasions (Ball et al., 2005) and again in this review. The situation has no doubt been '*exacerbated by a widespread and possibly misplaced fixation in this country with progressing to Higher Education (HE) rather than a fuller consideration of more apprentice based or vocational courses*' (Farmer, 2016, p. 28), but the issues are even more deep seated than this, and find their way yet again back to our cyclical economy. Whilst a redistribution of the grant funding from the CITB across those sectors where innovation is most likely to be embraced would be beneficial, this does not go to the root causes that have led to this lack of uptake in skills training. The SMEs still have to feel confident that the training is worth investing in, which they will not if they feel their individual risk taking will only benefit the collective whole.

This is emerging as a key issue – the need for the reward received to be directly related to the risk taken. Too often in a complex structure such as the housing industry, the very real risks taken by small individual companies when attempting to adopt new innovative solutions only show returns if others in the supply chain also agree to participate, and even then, those returns tend to be distributed disproportionately widely compared to where the individual risk was taken. It is these concerns that are possibly behind the reluctance or even inability of supply chain companies and SMEs to participate in the innovation process.

> *There is often an unfair burden placed lower in the supply chain to train. It is clear though that despite the financial risk, some parts of the supply chain have actively embraced long-term thinking around skills and training. This seems to be often led by a non-financially motivated sense of collective responsibility to do the right thing and maintain a sustainable level of resources for the industry in the future. Unfortunately, this is not the industry-wide approach and is not inherently scalable.*
> (Farmer, 2016, p. 28)

The impact of not addressing a fair sharing of risk is revisited in Part II where even governments are shown to be failing to reap the necessary rewards for their efforts that would enable them to continue to prepare the environment for entrepreneurial development (Lazonick & Mazzucato, 2013).

[Disproportionate individual risk for an uncertain and shared reward stifles the adoption of innovations]

7 Workforce Size and Demographics

This perhaps more than anything else shows the extent to which macroeconomic events can and will influence the way in which the housing industry has to evolve. There is nothing unforeseen about this looming crisis, although our departure from the EU has no doubt exacerbated the situation, at least in the short term:

> *Perhaps most worrying of all of the symptoms identified in this section, is the fact that the pure physical capacity of the construction industry to deliver for its clients appears to be in serious long-term decline.*
> (Farmer, 2016, p. 32)

The impact that an undersized workforce will have on any decision-making process regarding how to deliver housing will effectively nullify any other considerations in our short- to medium-term future. The government's planned expansion of our skilled labour pool is dwarfed by the replacement demands caused by retirement figures that have been apparent for over 20 years.

A combination of an ageing workforce, low levels of new and an overlay of deep and recurring recessions which induce accelerated shrinkage, now threatens the very sustainability of the industry. It is potentially in danger of becoming unfit for purpose.

(Farmer, 2016, p. 32)

Unlike what we have seen in the decline of many other UK industries, complete construction solutions cannot easily be supplemented by imports, resulting in a dependency on migrant labour, particularly in London. It is this decline in our pool of skilled labour, more than anything else MMC solutions have to offer, that is driving the push for off-site manufacturing. This is especially true in London and the South East, where completed units can be manufactured in regions of lower labour costs and brought in for erection (although it should be remembered that 'outside London' could also be overseas once the concept of modular housing has been proven and accepted).

The scale of this problem is reflected in the Farmer Review's lack of answers on this point beyond saying that the government's house-building targets '*must now surely be seen as increasingly unrealistic*' (Farmer, 2016, p. 33). In the longer term, it must be assumed that the numbers attracted into the industry will increase as the demand pushes up wages in the sector, but this leads on to what is already an inconvenient truth about the cost of our current housing solutions: The reason why the volume housebuilders are reluctant to increase their output is because they already supply enough – or, to be more accurate, just less than enough – housing for the market that can afford to buy their product. The country's more pressing need is for housing that is affordable below this price point, and that is what has to be delivered with a shrinking workforce that will demand higher wages to encourage it to grow. Basic macro-economics will not only dictate how we build, but what we build, and would suggest that it cannot be more of the same.

[Our shrinking workforce will not grow to service the demand until the housing we build is affordable by enough of those making up that demand]

8 Lack of Collaboration and Improvement Culture

'*The construction industry's "collaboration problem" is at the root of its change inertia*' (Farmer, 2016, p. 33). This comes back to the adversarial nature of business transactions and contractual agreements within the supply chain, and how they prevent the cross-fertilisation of ideas and knowledge. BIM provides both an analogy and a possible solution to this in that it requires all participants to be on board before its real benefits can be appreciated by any one company or by the project as a whole. Its presence as a data management tool also opens up lines of communication in a collaborative way which should then theoretically lead to further avenues of co-operation, but getting BIM established at the outset – and maintaining that collective ideal once it has been – is a major undertaking.

Past experiences of the off-site manufacturing industry have proven how important it is to gain a critical mass of industry players to provide it with a constant throughput of work. Towards the end of a boom time, however, when the speed of construction becomes less critical and the supply of on-site labour more available, the alliance begins to falter, often leading to the collapse of the off-site provider (Mann, 2017). This is potentially a cyclical inevitability of a boom and bust economy, unless the supply of work can be guaranteed during a downturn. For this there would seem to be three possible solutions. Either the government has to provide that need in the form of social housing demand (Lyons, 2014, p. 22), or the off-site industry has to diversify, or it must provide its own market for housing. The solution to this and any form of collaboration, where the aim is to reduce the level of future risk, is to have at least some control over the other participating businesses involved. The New Entrants' solution to this is to not rely on government policy decisions to protect their investments, but to effectively control elements of the supply chain from concept design through to sale of the finished property, or even beyond through long-term rentals. They are also, as a consequence of this, more able to introduce and demand the use of BIM throughout their supply chain, without fear of that chain breaking. The more achievable solution in their eyes is therefore to create an alternative simpler marketplace rather than attempting to correct the existing profoundly complex one.

> *The industry's route map to collaboration and high efficiency new delivery models can only be underpinned by BIM and the importance of its adoption cannot be overestimated.*
>
> (Farmer, 2016, p. 36)

The practical realities of creating a more collaborative environment are discussed in Part II to show how such chicken and egg challenges cannot be brought about internally. Again, the act of calling for change without recognising the reasons why this has not been successful on previous occasions runs the risk of yet again postponing those changes taking place.

[The task of reintroducing collaboration into a complex and adversarial marketplace is in itself a major barrier to change]

9 Lack of R&D and Investment in Innovation

The figures for investment in R&D in construction in the UK are lower than in any other sector.

> *This is symptomatic of a lack of interest in or incentive to consider modernisation in the industry despite meaningful tax offsets being offered.*
>
> (Farmer, 2016, p. 35)

Whether or not this is an over-simplification of the reasons behind the problem, it is, nonetheless, unhelpful when it comes to finding a solution. But would a more complete picture, taking into account the role of disproportionate risk, supply chain complexities, public conservatism – and all the perceptions and past experiences that have fed into that – provide us with anything more useful to work with?

> *This review heard evidence from businesses that are investing in innovation but appear to meet problems in getting new products and propositions to market at any scale. This is often due to a deep-seated perception of risk within the wider supply chain.*
>
> (Farmer, 2016, p. 35)

The review focuses at this point on the trials of off-site manufacturing and the many barriers to its acceptance. It recognises 'the cyclicality of construction demand', the negative perceptions from past experiences, the reluctance of investors, lenders and insurers to participate, the inappropriateness of our traditional tendering process, the need to co-ordinate supply and demand throughout the industry, the power of the traditional developer model over the marketplace, the inconsistency of government policies, plus all the factors already covered above that feed into the wall of resistance that needs to be overcome. But for all this, Farmer sees off-site manufacturing as the only viable solution to the extent that it will either be addressed here or it will be won by overseas markets, notably from Asia.

The underlying issue here is risk, and it is felt across the whole industry, not just by the small entrepreneurial companies, who are often portrayed as the only players gambling with their investments. Governments and large corporations also take risks by investing in creating the fertile environments for those start-ups to flourish, and the buying public takes risks in moving away from the accepted housing norms with possibly the biggest investment they will ever make. There is no avoiding the need for the whole industry to share this risk to break the stalemate of inertia that has settled upon it. For that to happen Farmer implicitly suggests that the industry needs an external influence to intervene.

> *Ultimately, innovation led modernisation continues to be inhibited at all levels by the lack of industry-wide strategic leadership with a more integrated client and industry agenda.*

(Farmer, 2016, p. 36)

If that is not to come from government, which is the reluctant conclusion that the New Entrants have come to, it is up to these industries operating on the periphery of housing to step in with alternative solutions that simplify and de-risk the offer, not just for themselves, but for the investors, the insurers and the end users.

[Perceptions of risk due to complexity and uncertainty stifles investment in innovation]

10 Poor Industry Image

These nine symptoms culminate in the industry's poor image, deemed worst within the housebuilding and residential sector, due to high risk, poor health and safety, austere working environments, embedded prejudices and, perhaps crucially, poor job security, reinforced by a feeling in general construction circles that it is this sector that is most at risk from 'boom and bust' due to a strong alignment with the housing for sale market. If this industry is going to attract the workforce needed to survive, let alone grow to deliver the government's targets, and do this from within the UK, the most influential factor that will dictate the way the housing industry will evolve, it is suggested here, will be whatever it is that appears most attractive to the labour market. And if high incomes is not an option – as these houses will have to be more affordable than those currently being built traditionally – it will most likely be down to work environment, 'high tech appeal' and job security. It may well be that the ground rules are so fundamentally different for this round of the cycle as to make the case for off-site manufacturing unstoppable, and the negative past experiences, for the first time, surmountable.

[Off-site manufacturing may become the only option for improving the industry's poor image, but would still need to deliver more stability through recessionary periods]

So What Now?

The purpose of that exercise was to drill down further into the Farmer Review's ten symptoms to try and establish what the shared root causes might be that lie behind them. Asking 'why?' until there is nowhere else to go takes time, and rarely leads to an answer that can be easily acted upon, because the root causes to most problems lie beyond the remit of the person, business, industry asking the question. But this does not negate the reasons for undertaking the exercise. It is an essential part of the process of defining what can be achieved within the constraints of what can be changed. But this is not an exercise limited to industry asking 'why' of government decision making. It is equally applicable to governments needing to ask 'why' of the construction industry in order to better understand what it is attempting to influence. The tool developed in Chapter 4 is therefore aimed at all parties to help create that two-way street of information that is currently not flowing freely in either direction.

Bibliography

Ball, M., Cowley, J., & Slaughter, J. (2005). *The labour needs of extra housing output: Can the housebuilding industry cope.* Home Builders Federation.

Baseley, S. (2016). 'Javid is wrong, we do not landbank'. *Inside Housing*, December. www.insidehousing.co.uk/comment/comment/javid-is-wrong-we-do-not-landbank-48863

Communities and Local Government Committee. (2017). *Capacity in the homebuilding industry*, April. UK Parliament.

Egan, J. (1998). *Rethinking construction: The report of the Construction Task Force.* Department of Trade and Industry, UK Government.

Farmer, M. (2016). *The Farmer Review of the UK Construction Labour Model.* Construction Leadership Council. www.constructionleadershipcouncil.co.uk/news/farmerreport/

Hansmann, R., Mieg, H. A., & Frischknecht, P. (2012). Principal sustainability components: Empirical analysis of synergies between the three pillars of sustainability. *International Journal of Sustainable Development & World Ecology, 19*(5), 451–459. https://doi.org/10.1080/13504509.2012.696220

HM Government. (2015). *Register of All-Party Parliamentary Groups*, September.

Iacono, R. (2018). The Nordic model of economic development and welfare: Recent developments and future prospects. *Intereconomics, 53*(4), 185–190. https://doi.org/10.1007/s10272-018-0747-2

Lazonick, W., & Mazzucato, M. (2013). The risk-reward nexus in the innovation-inequality relationship: Who takes the risks? Who gets the rewards? *Industrial and Corporate Change, 22*(4), 1093–1128. https://doi.org/10.1093/icc/dtt019

Lyons, M. (2014). *Mobilising across the nation to build the homes our children need.* The Lyons Housing Review.

Mann, B. W. (2017). Precast prepared for offsite's second coming. *Construction News*, 19–21.

Mann, W. (2017). Has offsite's time finally arrived? *CIOB*, October.

Roper, S., & Tapinos, E. (2016). Taking risks in the face of uncertainty: An exploratory analysis of green innovation. *Technological forecasting and Social Change*, *112*, 357–363. https://doi.org/10.1016/j.techfore.2016.07.037

Siebert, M. (2019). *Concrete and the barriers to innovation in UK housebuilding*. PhD thesis, University of Nottingham.

Toma, L., & Gheorghe, E. (1992). Equilibrium and disorder in human decision-making processes – some methodological aspects within the new paradigm. *Technological Forecasting and Social Change*, *41*(4), 401–422. https://doi.org/10.1016/0040-1625(92)90046-V

Turner, L. (2019). The role of SMEs in the UK construction industry. *RICS Construction Journal*, 2022. ww3.rics.org/uk/en/journals/construction-journal/the-role-of-smes-in-the-uk-construction-industry.html

4   The Methodology for Change

There has been so much written about change management now, much of it rebranding similar approaches and discrediting the many preceding theories that have been trialled over the years and across all possible disciplines. Change-Washing has been put forward as a term, meaning '*the process of introducing reforms that purport to bring about change but fail to result in any substantive shifts in systems, services or culture*' (Snow & Greenspoon, 2020). The danger is that this change management fatigue can easily be misconstrued as an unwillingness to adopt change itself, which is not necessarily the case. Recent incarnations of change management programmes have begun to recognise the limitations of top-down approaches, alongside the complexity and fluidity of real-life scenarios, but they still tend to be presented as quick-fix packaged solutions, which is the trap that this approach is keen to avoid.

Systems Thinking

So what exactly is Systems Thinking, why is it any different, and how can it be interpreted beneficially for the construction industry?

This may sound like a very subtle differentiator, but Systems Thinking merely defines ways for approaching problem solving, and in that respect it is very non-prescriptive. The very breadth of approaches available within System Thinking's arsenal is testament to that, and also to the maturity of a philosophy that has been evolving for over 50 years. The principles behind it are both straightforward and open to interpretation, which are important qualities in a situation where it is all too easy to over-complicate what is needed to bring about change, or to be more accurate, improve an industry's decision-making process so that it can bring about its own change more effectively. It is also an approach that recognises that, to be successful, it must be both simple to understand and simple to implement, because just as with any other innovation, the same hurdles of knowledge, motivation and ability need to be cleared to ensure its adoption.

Whilst Systems Thinking is itself a simple concept, over the many decades of its existence as a methodology it has evolved to cater for many different needs,

DOI: 10.1201/9781003332930-6

and it is interesting to map these various permutations against the sectors and hierarchical structures within the construction industry.

At its most basic, Systems Thinking is a holistic approach to analysis that focuses on the way that a system's – in this case the construction industry's – constituent parts interrelate and work together over time, often within other larger systems, in this case the broader economy. The term 'organism' is often used to describe these systems, working within larger systems representing the organism's environment. The analogy with nature is intentional, as the methodology reflects how the natural world operates, and the way in which its own systems work through the effects of reinforcing and balancing processes. A reinforcing process leads to the increase of some system component, which, if left unchecked by a balancing process, will eventually lead to a collapse. A balancing process on the other hand is one that tends to result in an equilibrium being reached.

Shifting Equilibria

Unlike in nature, however, not all man-made equilibria are universally desirable, as is the case with the construction industry's current status quo, and one of the challenges is that of shifting the 'construction industry system' from one state of equilibrium to another. By way of example:

A major barrier to the successful adoption of modern methods of construction is the reluctance of the volume housebuilders to transition to a new way of building that they see as being of only marginal benefit to their business model, thereby reducing the market for innovative products, and holding back progress across the whole industry. In reality, they have no fundamental problem with change as long as it does not impact detrimentally on their profit margins, which often makes the cost of transitioning from one system to another the most important barrier to overcome.

The drive to introduce 'large format blocks' should have been a relatively easy promotional exercise. The blocks were larger than standard concrete blocks, but proportionally lighter, making it faster to construct internal and non-loadbearing walls. The thermal value of the blocks was also greater, making it easier to comply with the new building regulations being introduced at the time. Three factors however prevented that easy transition. The first was an example of internal disunity, the second, industry complexity and the third external uncertainty, the three subdivisions of the 'ability' of a business to transition discussed in the previous chapter:

- In trials, the bricklayers, unknown to the company trialling the blocks, slowed down the speed at which they worked because they were used to being paid at a rate per block and thought they would be paid less for laying larger blocks at the same pace.
- The large format blocks were designed to optimise the health and safety weight limits for manual lifting, resulting in a dimension that

had no bearing on the industry's 900mm module, although only out by 30mm at 645mm instead of 675mm (three bricks' length). As this would have meant having to redesign entire pattern books of standard house types, the transition costs became too onerous.
- The fact that the blocks were an imported product from a limited supply chain that could deprioritise the UK in times of greater domestic demand meant that the risk levels were too great, especially after having changed designs to allow for their use.

None of these were insurmountable challenges, but none were foreseen, exemplifying the need for the holistic nature of a Systems Thinking approach. It is only by considering such issues from all perspectives and at all levels of engagement that a complete picture emerges of how different factors impact on each other, often in unexpected ways.

Using Systems Thinking's Methodologies

What this also exemplifies is the different 'flavours' of Systems Thinking, all of which have a role to play at different levels of interaction, from simple business-level decisions to the increasingly complex, 'pluralistic' and often 'coercive' environments that exist as the net of relevance is widened to encompass all the factors that impact upon the decision-making process.

The aim here is to understand those differences and apply them correctly, not to over-complicate the process unnecessarily. The construction industry has neither the time nor the inclination to learn about the inner working of Systems Thinking, but it can benefit from a process evolved from it that can be shown to be both relevant and appropriate in scale to any problem that arises. Bearing that in mind, the following explanation is not critical to understanding the process being developed, a process that to be successfully adopted must not only work, but work without requiring an unrealistic investment in time and effort to understand its benefits. So do you need to read the rest of this chapter? I would say 'yes', now I've written it, but I can also see why it might begin to feel like a journey into another world of little relevance. If it helps you decide, the next section is structured like this:

- First we explain the evolution of Systems Thinking as a methodology, and how it's grown to encompass increasingly complex 'systems' or real-life situations.
- Then we try to relate these different 'flavours' of Systems Thinking to the construction industry and show how they compare with currently used problem-solving approaches.
- With that out the way, we can then move on to explaining how an easy to use decision-making tool can be developed using these principles, but without having to expect the industry at large to get 'underneath the bonnet'.

- We do this by using a decision tree flow diagram that breaks the process down into some key questions that always need to be asked and showing how, if these are approached in the right order, the problem being confronted can sometimes be defined, if not resolved, a lot quicker than expected.
- And finally, we arrive at how to make a truly comparative assessment of the options available, when dealing with disparate and subjective information, unknown risks and constantly changing parameters.

By using industry examples throughout, the benefits of taking a more methodical approach to problem solving become clear, which prepares the way for Part II, where the process is then used to address some of the industry's more intractable problems, including the all-important challenge of implementation. If the approach being developed to create the tools we need is deemed too complex to use, it will have fallen at the first hurdle, and the complex problems these tools are intended to resolve will be all that remains.

The Evolution of Systems Thinking

Understanding how Systems Thinking can have a relevance to the construction industry at many different levels is nevertheless interesting to see. The rings in Figure 4.1 represent the boundaries that we draw, often unknowingly, around

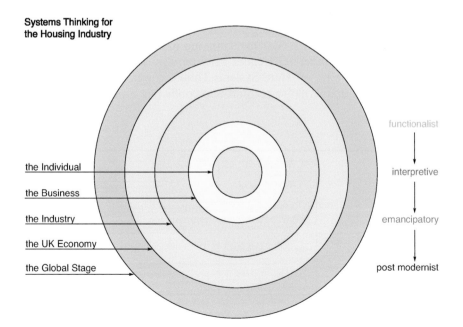

Figure 4.1 The different flavours of Systems Thinking.

the problems we're trying to solve – starting with 'it's all about me' in the centre. Systems Thinking defines these as boundaries between an 'organism' and the environment it exists within, and where that boundary is drawn depends on the organism being investigated. It might be a business within the industry, the industry as part of the UK economy or the UK economy on a global stage. Ultimately, the only stasis can be at a global level, as all other defined 'organisms' are interacting within an environment which is part of a greater organism – as climate change is now teaching us.

Treating organisms as black boxes where the outcome within those constraints is all that is fully understood is to deal with the consequences of a problem rather than the root causes that invariably lie further afield. By introducing the concept of reinforcing and balancing processes, or positive and negative feedback loops, those boundaries are crossed as other exogenous factors are brought into play. Understanding those feedback loops is the first step towards controlling them and being able to influence how an organism can be made to function more efficiently within its wider environment.

This is how Systems Thinking now operates, but it started out as a far simpler definition of cause and effect. The transition from a system responding to an already complex natural world to responding to a human environment was an evolutionary step for Systems Thinking as it introduced purposeful subsystems within the organism analogy whose responses were driven by a variety of different human motivations. This more advanced way of understanding real-world scenarios led to the introduction of terms to define these many roles – stakeholders, decision makers, actors, customers, witnesses, problem owners and problem solvers. In the UK housing industry for instance, home buyers are the customers, but instead of being stakeholders, have become witnesses, rarely able to influence the house-building process.

The application of these Systems Thinking methodologies to real-world problems highlights other differences between approaches that have since emerged. The first wave of 'Hard Systems Thinking' was recognised as being unable to deal with alternative realities and conflicting motivations, but was

			PARTICIPANTS			
			UNITARY	PLURALIST		COERCIVE
SYSTEMS THINKING			HARD ST	SOFT ST		
			machine learning	human decision making		
			TYPE A	TYPE B	TYPE C	TYPE D
			Functionalist	interpretive	emancipatory	post modernist
			clear goals, little dissent	subjectivity	unfairness	no consensus position
			consensus expected	consensus sought		compromise sought
SYSTEMS	SIMPLE	unified structure	Large, established business			
	COMPLEX	multiple sub-systems	Vertically Integrated business model		Construction Industry	Housing Industry

Figure 4.2 Systems Thinking categorisations.

well suited to large organisations with clear goals and little dissent, such as was found in the Space Programme, where Systems Thinking's parent concept, Systems Engineering, was first practised. The recognition of the need for a variety of approaches led to research into the 'System of Systems' methodology and the subdivision of systems into Simple or Complex and the participants into 'Unitary', 'Pluralist' and 'Coercive'. The housing industry, as we will discuss later, would fall into the Complex / Coercive category, with its multiple subsystems and a wealth of participants with alternative and often conflicting needs and opinions.

To take a slightly deeper dive:

The housing industry resides in this corner of the table because it is too adversarial even for Pluralist Soft Systems Thinking, where alternative perspectives are considered with an understanding that a consensus viewpoint can be reached across all stakeholders. In more 'coercive' environments, the last resort for Systems Thinking, seeking a compromise is the only option available. The further categorisations in the table define the approaches that have evolved over time to deal with these increasingly complex scenarios, ranging from (A) objective, to (B) subjective, to (C) discriminatory, to (D) coercive:

Type A approaches tend to be for Unitary problems and are mainly Hard Systems Thinking approaches.

Type B approaches are predominantly Soft Systems where the importance of subjectivity is recognised, and consensus can be sought.

Type C approaches are brought into play when there are signs of unfairness in the system, and some alternative viewpoints are not being recognised, or some stakeholders are being disenfranchised through discrimination of some kind. The aim, however, is still to find the consensus position, using 'Emancipatory Systems Thinking'.

Type D are 'Postmodern Systems' approaches that seek to promote diversity in problem resolution. There is a belief behind these systems that the coercive nature of many problems means that there can be no consensus position and any complete solution is therefore unlikely to exist. All that can be achieved is a compromise solution at best and, more likely, merely an understanding of why the solution achieved is unjust but inevitable.

Systems Thinking can therefore be defined and categorised in many different ways, but the aim is not to choose any one of them over another, but to understand and implement all of them when and where it is appropriate to do so. There is, nevertheless, a tendency to favour the more recent offerings as ones that reflect a more mature understanding of the complexities involved in decision making, which is certainly the temptation within the housing industry with its multiple and often diametrically opposed motivational drivers. Whilst this is understandable, not all scenarios are complex, and a methodology that can be applied in any circumstance needs to be adaptable to the situation in hand, and not require the situation to be shoehorned into the methodology's prescriptive way of working.

It is, nevertheless, the most recent incarnation of Systems Thinking, the Postmodernist viewpoint (Type D), that would seem to best represent the current state of the housing industry that we are focusing on here. This incarnation of Systems Thinking moves even further away from the certainty that defined the earlier models, into a space where the very purpose of prediction and long-term planning – and certainly five-year plans – are dismissed as pointless, even dangerous. In this respect, it is worth noting the political angle to this shift, as a move against predictability could be construed as a reinforcement of free market economics and against interventionist policies. There is however a counter-consensus of sorts that has emerged from this progression, suggesting that well-directed incremental changes that aim towards a predicted future position as it currently presents itself is the most effective way forwards, and more appropriate than radical shifts in policy that set a fixed course for what is in most cases a moving target.

So, having explained the evolutionary stages of System Thinking's development, is there a time and a place for each paradigm from Type A to D, from pure functionalism through to the near anarchic beliefs of the postmodernists? And if so, how can any of this be realistically integrated into an industry where the concept of R&D barely extends beyond a site visit?

How Systems Thinking Relates to Construction

The first point to make is probably that the construction industry is not intrinsically different to any other industry, but it is an increasingly complex and fragmented industry, and in addition to this, the housing sector is constrained by a powerful status quo in how it considers its options going forwards – which would suggest it could and should be benefitting from a Systems Thinking approach in some form or another. The missing piece of the change process is, and has been for some time now, the implementation stage, which is as true for a new way of thinking as it is for a new way of building. In other words, before using a Systems Thinking approach to help implement new strategies for building, we need to use it to implement a new strategy for problem solving, i.e. itself.

How to translate a methodology for complex decision making into a simple and relevant process where the benefit can be clearly seen is therefore the next task – one of deconstructing the problems discussed around how the industry currently operates to come up with an approach that deals with those problems, rather than pretending they are either peripheral and to be ignored, or that they will somehow cease to be problems once the transition to a new way of working has magically taken place.

The following are the necessary steps that will get us to that useable formula, starting with a clarification of what roles the different Systems Thinking paradigms need to play:

Looking at the different flavours of Systems Thinking, there would seem to be a strong association between each one and the point at which the boundaries to the issues being addressed are drawn, certainly with respect to how the construction industry operates. At an individual business level, for instance, the immediate concerns tend to be relatively simplistic and related to short-term targets, efficiency levels and bottom-line profit, and especially so for well-established mainstream businesses. In these instances, with little internal contradiction in play, the 'Hard Systems Thinking' functionalist methodologies provide a sound starting point, at least for understanding what that business needs in order to remain profitable. But in order to capture the impact that competing businesses might have on any strategies put forward at this level, or how a business's customer base might respond to the adoption of an innovation, a more 'pluralist' approach would be beneficial. This would be appropriate if looking at competing supply chain businesses for instance, and takes us into the realms of 'Soft Systems Thinking'.

Moving outside the immediate confines of the supplier and its customer base, the housing industry represents a broader boundary that includes the many other sectors whose input also needs to be considered – the legislative bodies, funding bodies, the various housing provider business models, and the design industry, in itself a multi-faceted sub-industry. To deal with these stakeholders' conflicting motivations requires a more nuanced paradigm that considers how consensus and collaboration can be achieved where siloed thinking and adversarial business practices are currently the norm. The aim here is to bring about fairer and more enlightened solutions, whilst still acting within the confines of the social and political environment in which the entire industry has to operate.

The next step beyond this pluralist approach accepts that there is not always a consensus viewpoint that can be reached, and the further the boundary is stretched to include this wider economic model, the more likely it is that alternative viewpoints and motivations will defeat the desire to resolve conflict and find consensus. This is where some voices begin to get drowned out, and where the levels of complexity that must be dealt with often constitute the 'Wicked Problems' mentioned earlier that cannot be solved, only resolved, based on the facts as they present themselves at any one time, and from multiple subjective perspectives. The 'coercive' models that have been developed to recognise this reality attempt to give a voice to those bodies, often defined as witnesses rather than participants, so as to rebalance the equation. In the housing industry, the buying public are often seen as little more than 'pawns in the game', certainly whilst there is an endemic shortage of housing, but even the government is looking increasingly powerless in the face of an ever more entrenched speculative housing business model that runs counter to the government's goals.

The Meaning of Wicked Problems

Rittel and Webber suggest that 'we are all beginning to realize that one of the most intractable problems is that of defining problems' (Rittel & Webber, 1973, p. 159). They go on to suggest that 'the social professions were misled somewhere along the line into assuming they could be applied scientists – that they could solve problems in the ways scientists can solve their sorts of problems' (p. 160). Whereas, in reality, 'societal problems … are inherently different from the problems that scientists … deal with. Planning problems are inherently wicked … Social [wicked] problems are never solved. At best they are only re-solved – over and over again' (p. 160).

These statements, although made nearly 50 years ago, seem to encapsulate the problems that have ensnared the UK's housing industry for almost as long. They also lend weight to the approach of focusing on the questions to be asked rather than on providing definitive answers. If the housing industry represents a wicked problem, the solution lies in defining that problem in a way that it can be re-solved at any time now and in the future with greater clarity than is currently the case.

The model in Figure 4.3 categorises the three levels of operational control that reflect the difference between how simple structures can be organised

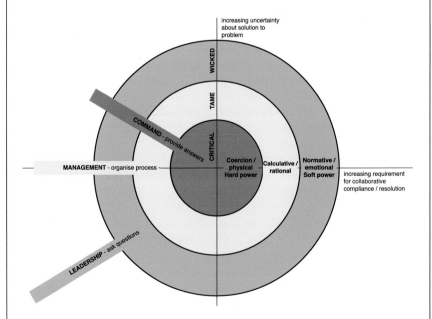

Figure 4.3 Critical, Tame and Wicked Problems.

through a command approach and how this needs to be replaced in the con-struction industry with a methodology that better reflects its complexities and uncertainties, and instead encourages a questions-based approach to problem solving.

Perhaps the inevitable final step for Systems Thinking beyond this admission that not all conflict can be 'managed' out of the equation is the world view that questions the very nature of discourse and what constitutes an acceptable window for debate. As this process of boundary expansion reaches its global end game, the criticism of idealism distancing the whole purpose of the movement from its original practical roots grows in strength. But without the honesty that the postmodernist thinking brings to the equation, the equally valid counter-claim is that the answers that the systemic and critical permutations of Systems Thinking offer are blinkered in their failure to recognise the broader context of the environment in which we must all operate.

Understanding the role that each paradigm was developed to perform is clearly an important part of this process, as is recognising how the housing industry is structured within a hierarchy of ever more heterogeneous boundaries. Only with this knowledge can the appropriate methods be brought to bear on the problems being confronted, and a necessary compromise be sought between the practicality of short-term survival and the ideology of a longer-term purpose. Ultimately, this is an exercise in being honest about what a person, a business, an industry, a socio-political economy – a civilisation – expects to achieve, and how best to go about satisfying those aims.

Radical Incrementalism

The term 'Trojan Mice' is often used to describe an under the radar grass-roots approach to change where small 'safe to fail' experiments can be used to trial strategies without generating too much 'change management resistance', but the term Trojan Mice would suggest that someone is being kept in the dark. Is that the management or are the changes being surreptitiously drip fed from above? In reality, everyone needs to be in agreement with any approach being adopted for it to succeed, and everyone needs to feel a degree of ownership. The approach developed here to reflect this is a policy of thinking big but acting small – to be aware of the end game or the radical vision, but to avoid making radical changes to achieve that where small, well-directed incremental changes might offer a better, safer, more productive route forwards. This is Radical Incrementalism, the difference being that the small steps being taken should be informed and directed towards the same agreed goal, but that goal should be constantly reassessed in the light of the unknown future environment in which we all have to operate. And it is that constant reassessment that requires a question-based approach that captures the needs of all stakeholders.

Developing the Decision-Making Tool

Throughout the book so far we have alluded to a number of important elements that need to be incorporated into this process for it to stand a chance of being adopted, all of which stem from taking a Systems Thinking approach where all perspectives are considered:

- All stakeholders must benefit from an innovation in a way that makes sense to them.
- The benefits to be accrued must outweigh the barriers to be overcome.
- A hierarchy of importance must be established as a way of weighting these pros and cons.
- This must also recognise the hierarchy of stakeholders in the decision-making process.
- All barriers to innovation must be dealt with by either removing or avoiding them, or even by just accepting them – but never ignoring them.
- There are three fundamental barriers to overcome – knowledge, motivation and ability – before any progress can be made.
- Knowledge of an innovation requires it to be translated into a language that will be understood; motivation to engage with it requires it to be viable; and an ability to adopt it requires it to be feasible.
- For many, perception is reality, and can still represent a barrier that must be addressed.
- It is the root causes that must be found and dealt with, not just their consequences.
- Asking 'why?' until there are no further answers to give leads to those root causes.
- Most root causes lie outside the framework of the problem being confronted.
- The framework of acceptable discourse should not be allowed to limit the questions being asked.
- Research into the 'unknowns' must have depth into the detail, length into past experience and breadth across all stakeholders.
- The path of least resistance that remains will be where the collective benefits of adoption are greatest and the barriers are least.

The Question-Based Approach

What is needed is a question-based approach that encapsulates these elements, but is simple to follow, proportionate to the problem in hand, and able to respond to rapidly changing circumstances. It should leave the user free to find their own answers based on knowledge gathered from all perspectives. It should be one that gets beneath the surface and back to first principles where necessary, and captures past experience as well as current know-how. If such an approach can be developed, it will result in better, faster, more engaged decision making, to the benefit of everyone involved.

Before starting to build that question-based framework, it's important to recognise that the parameters listed above that an innovation must comply with to optimise its chances of success are equally applicable to this exercise. Promoting a new way to approach problem solving is no different to promoting any other innovation, and must overcome all the same barriers to adoption.

Problem solving at its simplest is a four-stage process:

Defining the question
⬇
Researching the unknowns
⬇
Resolving the problem
⬇
Enacting the solution

Irrespective of who has the problem to be solved – either the promoter of an innovation or the potential customer looking to adopt it – this can be translated into four fundamental questions:

What are the benefits of an innovation?
•
What are the barriers preventing its adoption?
•
Do the benefits outweigh the barriers?
•
How can those barriers be either removed or avoided?

The real challenge, however, comes in making this process quick and simple enough to be considered by a time-constrained industry without short circuiting the process and rendering it either unreliable or inaccurate. A fixed and exhaustive procedure can often result in an excessive workload for simpler tasks, and quickly lose its appeal, but designing a process that has the flexibility to always be 'proportionate to the problem in hand' involves recognising the difference between 'short cuts' and 'short circuits'. Weighing up the pros and cons of an innovative solution against an existing way of working, and doing this from multiple perspectives, is not a straightforward process, and is where the necessary decision-making processes often get short circuited, resulting in poor outcomes. A relatively safe and reliable short cut, however, comes in the order in which the questions are asked, starting with the easiest to answer before moving onto ones that will require a more complex comparative analysis, but only if and when that next step becomes necessary.

The Decision-Making Tree

The following flow chart defines such an approach and will form the framework of a decision-making tool that guides the user through a series of questions that aim to ensure all the relevant information is captured, but no more than is needed to reach an informed conclusion. The full meaning of these steps is explained next, but the order is key as sometimes only the first of these six questions will be enough to establish the outcome of the process.

Check for unresolvable non-negotiables

⬇

Establish the hierarchy of each sector's influence

⬇

Look for mutual benefits to assess appropriateness

⬇

Check for exclusivity to define market dominance

⬇

Assess viability of the market defined

⬇

Assess feasibility in terms of internal, industry and external risk

Working through this decision-tree flow diagram begins to show how these questions can help to reduce the time spent on unnecessary investigation by ruling out untenable options at the outset. If, for instance, the question was whether to build a house traditionally from brick and block, or to use a certain off-site volumetric timber frame company, researching the pros and cons of both approaches could become a lengthy process. By first asking if there were any key motivating factors that would overrule all other considerations, the local authority client announces that using local labour is one of its key objectives, and that no further evidence for or against will change this decision (Figure 4.4).

The reasoning behind each of these six steps and the order they are best approached in is explained here, together with example scenarios for how each step can reduce the scope or even negate the need for the next to be taken. The examples used here are all relating to the use of concrete products and the ways in which the concrete industry has tried to promote them:

Check for unresolvable non-negotiables

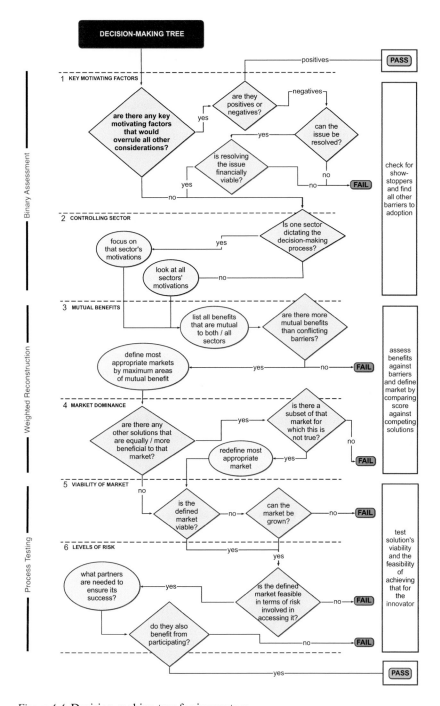

Figure 4.4 Decision-making tree for innovators.

Key Motivating Factors

Many decisions are made based upon one 'key motivating factor', i.e. a factor that is, for whatever reason, not up for discussion. This may be due to a client's demand, a planning requirement, or any one of a multitude of 'non-negotiables', but it is where the investigation should begin. If an 'unresolvable non-negotiable' is found to exist, the process ends there, although many non-negotiables are easily resolved once understood, and therefore just become factors to be considered alongside all the others, in which case the process moves on.

Primary question:

What is the key motivating factor that is dictating the decisions being made?

Secondary questions:

Are these internal decisions or dictated / influenced by other stakeholders?
If so, which stakeholders?

Example:

A development won an environmental award for its use of thermal mass and natural ventilation, and the developers used this heavily in their promotional literature. When asked by the concrete industry why they had decided to build a massive concrete structure on a tight city site, the developers admitted that they had had no choice due to the high water table. Their structural engineer had calculated that due to the need for two stories of underground carparking, no other method of construction would have been heavy enough to stay in the ground. The decision to use concrete was therefore dictated by site conditions and was not one taken after researching the benefits of thermal mass in passively reducing energy use.

Establish the hierarchy of each sector's influence

The Dominant Sector

The next question is how to gauge which sectors are most likely to influence the final decision. In a similar way to how the unresolvable non-negotiable factor can short cut the process, so too can one sector dictating the choices made. This therefore provides a second rule to minimise the time taken investigating a proposal: Before trying to prioritise the importance of any one factor being considered, prioritise the importance of the sectors involved in the decision-making process. If one sector is controlling that process, less emphasis should be given to the weight of factors from the other sectors' perspectives, unless of course the intention is to rebalance the playing field.

Primary question:

Which sector has most influence over the decisions about what to build and how?

Secondary questions:

How much, if any, influence do you have over that sector?
How much, if any, influence do those other sectors have over you?
How much are your decisions influenced by financial, social or environmental concerns?

Example:

When the concrete industry tries to promote the use of concrete as a building material in housing it talks to architects about the benefits of thermal mass in controlling internal temperatures. The principle is well understood, but until the SAP (Standard Assessment Procedure) software used to calculate the heat loss of a building measures the dynamic impact of heat transfer through a structure, the benefits of thermal mass will not be realised on paper, and will therefore rarely be implemented. In this instance the dominant sector is the legislative body controlling the regulations that must be complied with. Until this is dealt with, there is little point in promoting the benefits of thermal mass elsewhere.

Look for mutual benefits to assess appropriateness

Mutual Benefits

Now that the field of factors that need to be considered has been tested and most likely reduced by the first two questions, the comparative analysis begins: 'Do the mutual benefits outweigh the conflicting interests?' The aim of this question should be to seek out scenarios where the innovation being proposed is of mutual benefit to as many stakeholders as possible. The more synergies that can be found, the more appropriate the solution is for the market being approached. If the barriers to adoption outweigh the benefits, that is a clear sign of an inappropriate market, but it is not just a case of numbers, as not all benefits or barriers carry similar weight. The most fertile ground for an innovation however will always be that where there are the least barriers or no barriers at all to its adoption, and where there are many beneficial criteria from many stakeholders' perspectives.

Primary question:

What are the mutual benefits that arise from the use of this solution?

Secondary questions:

In which markets are these most pronounced?
What are the barriers from the client's perspective?
What are the barriers from the supplier's perspective?
Can any of these barriers be overcome?
Are the remaining barriers outweighed by the mutual benefits?

Example:

The Hockerton Housing Project has had national exposure for over 30 years as an innovative site of geoarchitecture, where partially buried south-facing homes have been built to prove the benefits of passive solar gain and the thermal mass benefits of the ground as an insulator. Despite decades of promotional endeavour, however, there has been relatively little interest from the mainstream housing industry in this as a scalable solution. For this to be of genuine interest, there has to be a mutual benefit, one that goes beyond a desire to reduce carbon emissions. That benefit comes from limiting the solution to south-facing sloping sites where the additional cost of building, calculated at 1% per degree of slope, can be negated by adopting a cut and fill approach on unviably steep sites, meaning no soil needs to be either delivered or removed from site to create the earth bund needed for partial burial.

Check for exclusivity to define market dominance

Exclusivity

In reality, an exclusive market where one solution is a clear favourite is rare, but refining the selection criteria until this is the case eventually results in the most appropriate, and thereby the most viable, market for a product, and does help define where a new product would have the best chance of being adopted and gain a foothold from which it could expand.

Primary question:

Are there any other solutions that are equally or more suited to the market being pursued?

Secondary questions:

Can the market be redefined so that this is not the case?
Is enough known about the alternative to ensure this is a true competitor?

Do other sectors' influence impact on this comparison?

Example:

Custom self-build is a market that struggles to find viable methods of construction that suit its customers' exacting requirements – flexibility, individuality, simplicity, affordability… All too often, groups of self-builders fail to get their projects off the ground due to timing issues with so many different parties having to agree, finance and progress the site works needed before building can commence. Those that do then have to contend with boundary issues as each party tries to develop their own site independently and at a different pace. One solution that the concrete industry could exploit in this market with absolute exclusivity would be to provide a serviced site with a concrete crosswall shell for a housing development based on a planning agreement to maximise a site within an agreed envelope. All perimeter and boundary walls, floor and roof heights would be dealt with in a simple contract, beyond which the self-builders would be free to build out their individual plots at their own pace, and without fear of weather damaging the initial structure.

Assess viability of the market defined

Viability Check

By this stage, the market being defined may be the most appropriate, but due to the many criteria being applied to achieve this, it may also have become very niche. There are many ways in which viability can be measured, and now that the optimum market being considered has been fully defined, all other factors relating to viability – from all perspectives – have to be incorporated into the equation; not just the size of the market, but its growth potential, its accessibility, future stability in the face of economic downturns or competing providers, its impact on a supplier's existing product markets and the immediacy of returns to be expected.

Primary question:

Is the defined market viable from the supplier's perspective?

Secondary questions:

Does it represent an increase in product sold when offset against the replaced solution?
Is the market accessible and secure in its own right?
Is that new market under threat from competition elsewhere?
Is it a growth market with support from other sectors?

Example:

Kerkstoel is a Belgian family-owned company producing twin-wall concrete panels that has supplied their product into London for commercial projects over the past 20 years. When a UK company attempted to buy them, only to find that Kerkstoel was not for sale, they decided to set up their own UK business in competition. Not only was Antwerp closer to London than their operation in the Midlands, the market for twin-wall components was not viable outside of the London area, which Kerkstoel had calculated to be the only market worth supplying to in the UK. Establishing the viability of a market avoids developing solutions that are then left looking for problems to solve.

Assess feasibility in terms of internal, industry and external risk

Feasibility

Feasibility is a measure of practicality, and has been defined here in terms of the level of co-operation or 'buy-in' needed from others for an innovation to be successful and the levels of risk that entails for all those involved. At this juncture the aim is not to measure that risk, but to understand where it emanates from so that it can be either removed or avoided to prevent it becoming the immovable barrier that prevents progress being made. Risk, whether it emanates from dis-unity, complexity or uncertainty, represents an unknown variable that needs to be removed from the equation before that equation can be fully resolved.

Primary question:

Is the defined market feasible in terms of the risks to both the innovator and the adopter?

Secondary questions:

Internal risk: Do all at a company level understand the implications of the pro-posal and agree to its adoption?
Industry-level risk: Can the industry-level risks be reduced by more collaboration between the stakeholders involved?
External risk: Has the societal and political environment been factored into the client's financial equation?

Example:

Passive solar design requires housing to be designed and built to utilise natural ventilation and maximise solar gain, with thermal mass factored in to absorb and store this free heat. Proponents of this solution are evangelical about its potential, but fail to recognise its limitations in terms of scalability. Not all properties can be orientated to have large south-facing windows, so this can never be a total solution, only a partial one. Immediately this reduces its appeal by introducing the need for a split approach. Other feasibility issues that add to the risk associated with passive solar design are the uncertainty over future climate change induced weather conditions, and the fact that it is an inherently passive solution, with little control over the outcomes. Add to this the need for upfront design decisions to be taken and adhered to for a passive building to function correctly, and it can be seen why the MEV (Mechanical Extract Ventilation) industry is so often relied upon to 'fix' buildings with mechanical interventions later down the line.

Weighting the Factors and Measuring Risk

This linear process provides a useful framework for asking the right questions in the right order, but does not as yet help establish what those questions are beyond some top-level categorisations. Within each of these stages of analysis there are still many decisions to be made regarding the balance of benefits to be realised against the barriers to be overcome in achieving them. How, for instance, can the familiarity with an existing method of construction, or a supply chain or a workforce, be measured against the benefits to be gained from adopting an entirely new but supposedly superior solution? In other words, **how can risk, by definition an unknown quantity, be measured?**

This is the crux of the problem faced by the construction industry, and ultimately by those wanting to transform it. Uncalculated risk is the barrier to change, and rather than being confronted, it is being ignored because of the difficulties in measuring it, or dismissed as being peripheral to the argument. The aim now, therefore, is to find a way of measuring these unknown variables in as simple a way as possible. Normally this would involve complex quantitative and comparative assessments, with weighted scores given to definable factors thought to be relevant to the decision-making process, but the interrelationships between these factors together with the subjectivity of what these benefits actually represent make this an impossible task to carry out reliably.

The approach taken here therefore is more of a qualitative assessment where we just continue to ask 'why?' until the choices left become straightforward either / or decisions.

By continuing to interrogate the four top-level questions,

1   What are the benefits of the innovation?
2   What are the barriers preventing its adoption?
3   Do the benefits outweigh the barriers?
4   How can those barriers be either removed or avoided?

until they become binary decisions that can be answered in turn without the need for provisos, caveats or further investigation, the choices to be made become far simpler and less dependent on other factors. How well this works will become clearer later through example scenarios, but for now, it's the subdivision of the questions framework that needs to be developed, based on the statements listed above, which in turn were derived from the questions posed in the previous chapters.

1 Defining the Benefits

Who benefits from an innovation? Benefits can be broad, but tend to be measured primarily through financial viability, for (1) the innovator, for (2) their customers and potentially for (3) society as a whole. But that viability means

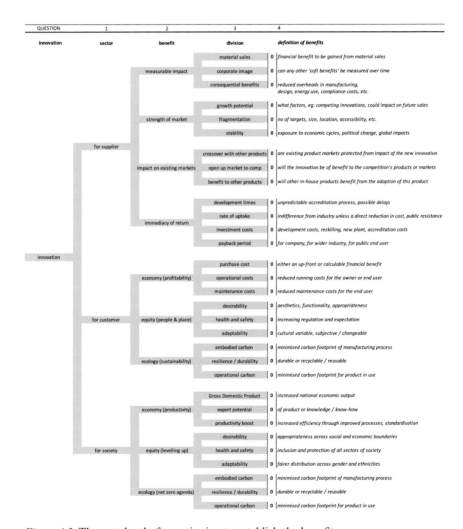

Figure 4.5 The next level of questioning to establish the benefits.

more than just bottom-line profit. At a societal level, for instance, the financial benefit may come indirectly through improved health outcomes, and even at a company level that benefit is rarely immediate and can be dependent on payback period calculations to offset initial outlays, energy saving in use, etc.

The divisions in Figure 4.5 have been created under each of these three subcategories to generate further levels of questioning. The expectation is not that each of these areas of interrogation will always be relevant, but that no relevant question that could be asked will fall outside of these categories. In that respect, this is not a finite list, but something that can evolve, grow or be personalised should that prove necessary.

By way of example, suppose the concrete industry were to suggest promoting piling as an alternative foundation system because it uses less concrete and would therefore improve the sector's sustainability image. For them it would mean less concrete sold but the promotion of a higher-value product that would compensate for this financially. But under this section:

Supplier > Impact on Existing Markets > Open Up Market to Competition > *will the innovation be of benefit to the competition's products or markets?*

the answer would be 'yes' as piling can be either concrete or steel, whilst the conventional foundation market is one where the concrete industry has no competition. This and many other factors need to be considered and the process being developed here aims to ask the questions that will ensure they all get included in the equation.

2 Defining the Barriers

In comparison, the barriers to innovation tend to be related to the more practical aspects of adoption, or the feasibility of changing working practices. By definition, these arguments tend to come from the individual businesses being encouraged to adopt an innovation, and stem from a far more critical understanding of the processes involved.

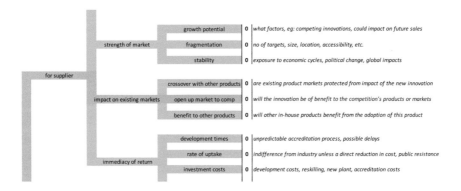

Figure 4.6 Example of benefits for the supplier.

Using the five industry sector divisions established earlier, together with the three hurdles of 'knowledge, motivation and ability' and the subdivisions found to exist within these, we have a starting point for capturing all the reasons why new innovative solutions might be harder to engage with than the benefits they are intended to bring might suggest. As with the Benefits table, this gives us a 'pigeonhole' system for locating any issue that's raised from any sector, with the functionality of being able to add further categories or to subdivide existing ones if the question asked still leads to an overly complex answer.

3 Do the Benefits Outweigh the Barriers?

We've now moved on from breaking down or deconstructing the problem into an array of binary decision points and now need to start constructing a case for or against the solution being proposed, using the decision tree flow diagram to ask the salient questions that will resolve the issue as simply as possible. Using a test case again helps to explain this process and the importance of treating these as 'Wicked Problems'. The example used here is a building material that is relatively new to the UK but in some quarters is being heralded as the new concrete and the future of sustainable construction – CLT, Compressed Laminated Timber. There is little point discussing the pros and cons of CLT in isolation, but equally, any alternative used as a comparator has to be related to the site being discussed, or the project, or the client's requirements from the perspective of the building being designed. There are in fact so many factors that may or may not be relevant to the decision being made whether or not to use CLT that the outcome could be different on every occasion – even for the exact same project a month later. Wicked Problems are problems that cannot be solved, only re-solved at the time of asking, and for that specific set of circumstances. It is for this reason that a question-based approach to resolving these kinds of problems makes so much more sense than trying to make a case for one option over another and then applying it to all eventualities.

Understanding the alternative perspectives is critical to this process, and the following argument is used here to demonstrate the complexity and fluidity of the decisions we need to make, based on the choice of CLT over concrete frame for a specific development in Hackney, London.

3a The Material Supplier's Perspective

The CLT Industry – London represents the region where CLT is most likely to be successful due to: it being an imported product; the tighter GLA (Greater London Authority) regulations on carbon emissions; and the restrictions on construction with respect to noise, pollution and access. Hackney, due to Hackney Borough Council's (HBC's) decision in 2012 to announce a 'timber-first policy' (Institute of Chartered Foresters, 2012), was particularly important to CLT establishing a foothold in the UK market.

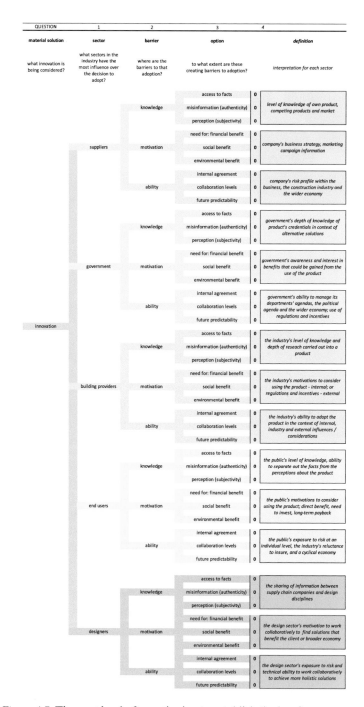

QUESTION	1	2	3	4
material solution	sector	barrier	option	definition
what innovation is being considered?	what sectors in the industry have the most influence over the decision to adopt?	where are the barriers to that adoption?	to what extent are these creating barriers to adoption?	interpretation for each sector

Figure 4.7 The next level of questioning to establish the barriers.

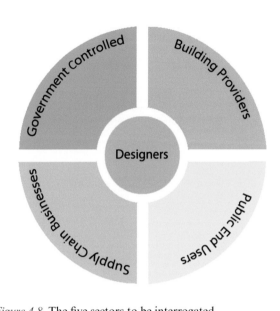

Figure 4.8 The five sectors to be interrogated.

The Concrete Industry – Normally cast in the role of disrupter, the concrete industry now finds itself protecting its established markets against a new entrant challenging its core market. On paper, CLT's credentials are strong when compared against concrete's on almost every front, leaving the fear of fire and water damage as the industry's only clear lines of attack.

3b The Government's Perspective

Keen to broaden the construction industry's supply base and reduce dependence on a few providers, the government would like to promote CLT, but cannot be seen to give too much support to any one sector of UK industry without being called to account by that sector's competition. Using sustainability as a reason for promoting CLT has already resulted in HBC having to reword its timber-first policy to reflect the lack of evidence that exists in support of timber's carbon credentials (Chartered Institute of Building, 2012). As an imported solution, however, the government's long-term interest would be in this becoming a UK manufactured product, something that, as will become clear later, is unlikely to ever materialise.

3c The Developer's Perspective

In this instance, the decision to use CLT had nothing to do with HBC's preference for timber, but was dictated by the moratorium that existed at the time on deep piling until Crossrail's route through Hackney had been decided. Using a lighter construction material avoided the need for piling and allowed the build to continue as scheduled.

3d The End User's Perspective

The public's, if not the client's, fear of timber construction due to the perceived threat of fire is largely misplaced, as the prevalence of timber construction fires happens during construction, not after completion (Lane, 2010). For many, however, perception is reality, and cannot be dismissed. Tightened regulations post-Grenfell that have resulted in financial ruin for many thousands of high-rise home owners left in negative equity is testament to this (Fire Protection Association, 2020).

3e The Designer's Perspective

The perspectives within this sector range from architectural practices looking for innovative solutions to champion, to a warranty industry reluctant to take unnecessary risks by stepping outside the proven norms. The likelihood of CLT ever being promoted by the design sector as a mainstream solution remains slim due to the question mark over financial security as an imported

Benefits of CLT over Concrete			
credential	known benefits	unkowns: disputed (lack of knowledge)	known barriers
adoption (early)	timber-first policies		not a UK product investment costs OSM barriers public perception of fire risk public perception of rot MMC warranties supply chain risk
material (current)	80% lighter safer, cleaner quieter on site faster on site (inc. second fix) more accurate (CNC cutting) flexibility of design error correction routing of services airtightness exposed finishes breathable construction weather independent	carbon emissions fire resistance glue / VOC	maximum height of build (less thermal mass) (acoustic performance) needs external insulation (Grenfell) above ground only structural span of 8m 3m panel width
potential (future)	UK production economies of scale	longevity proximity security of supply	catastrophic event reassessment of environmental impact

Figure 4.9 Table of pros and cons for CLT.

product, and has so far proven to be an unviable proposition as a UK manufacturing enterprise.

This is a snapshot of the pros and cons around the use of CLT, but serves to illustrate the uniqueness of every situation and the influence that individual scenarios can have over the decision made. There are two key unanswered questions, however, that will decide the future of CLT as a mainstream building solution: the future tightening of fire regulations and the reasons why CLT is unlikely to ever be manufactured in the UK. The second of those two questions has already been discussed in relation to anther imported product, Twin Wall where it was explained why sometimes importing products for the construction industry can be more profitable than manufacturing them here. As for fire regulations, with or without them, public opinion may well prove to be the deciding factor, overriding any statistical evidence to the contrary.

Armed with these alternative perspectives on the pros and cons of CLT and a far more holistic picture of the factors that should be taken into account, we can now return to the six steps outlined in the decision-making tree to see how asking these questions in the right order might shorten the exercise:

Key Motivating Factor (1): It is worth bearing in mind that these key motivating factors are often only of relevance to one stakeholder, but if they represent non-negotiable barriers, as Crossrail clearly did in this case, the level of influence held by that party becomes an irrelevance. So the simple choice to be made here was between using CLT or delaying the project until the path of Crossrail had been decided. Since waiting for that decision may still have resulted in having to use CLT, the cost of delaying the build programme was all that had to be calculated to inform the decision, which was therefore an easy decision to make – even if Crossrail were to eventually take a different route, which it did.

From this point on there would have been no point in the concrete industry pursuing this project, or in anyone debating the other pros and cons that still exist, as they would never be able to influence the outcome. Had the project come online a few years later, of course, that key motivating factor would have no longer even been a factor, and the process would have moved on to the *Controlling Sector Question (2)*. The outcome however would have been the same, as HBC's timber-first policy would have dictated the decision at a planning level. Two years later, that policy was contested and rescinded, opening the way for a more balanced discussion about the pros and cons of timber over concrete and their *Mutual Benefits (3)*. On balance, however, CLT still looks like a favourable option because of the proximity of the ports for delivery from abroad, the restrictions on noise and pollution in an inner-city residential area, and the local knowledge that had by now grown up around the use of CLT as a construction material – all of which are location-based factors.

Post-Grenfell, were the project to come online now, that decision may well be different again, with CLT's *Market Dominance (4)* potentially being restricted to buildings of a lower height, reducing the *Viability (5)* of CLT as a mainstream solution by squeezing the size of its market and increasing the *Risk (6)* associated with investing in its future.

So far this has been an entirely qualitative discussion requiring no complex comparative analysis of benefits, but with five separate time-related interventions potentially influencing the outcome. This is not to say that comparative costings will never be required, but there are many scenarios where conclusions can be reached without them, and when they are necessary, they can be narrowed down to far simpler comparisons based on near-binary choices. What does need to be calculated can then be clearly defined, which in this case fundamentally boiled down to:

The cost of delaying the build programme
vs.
The comparable costs of CLT over concrete frame

And to establish those comparable costs, the subsidiary calculations that needed to feed into that were:

- **The difference in foundation costs due to the reduced building weight**
- **The comparable levy payable to the GLA's Carbon Offset Levy**
- **The embodied and operational carbon figures used to calculate that levy**

As with the initial flow chart questions, and the relevance of the order in which they should be tackled, the same applies to the order in which these calculations should be approached, starting with the easiest to ascertain, which is probably how the GLA's Carbon Offset Levy is calculated. This is a recurring principle for reducing the quantity of research undertaken without reducing its quality. Is the levy great enough to influence the decision being made, or is it just an incidental fee to factor in to the costs? If it's not a defining factor, the research goes deeper, and also relevant may then be the cost implications of, for instance, the comparative time on site, lead-in times, health and safety provision and public safety measures, site access, impact on other design aspects – mechanical and electrical, fire, cladding systems, etc. – but again, the scale of impact of these in relation to the major factors above should be understood before deciding whether or not these calculations ever need to be carried out more accurately. The greater the difference found at the first levels of investigation, the less accurate the next level of research needs to be.

In case it had gone unnoticed, all of the three factors above are favourable to CLT, making the case for concrete actually seem extremely weak, but this only goes to highlight the power of the status quo – known systems, trusted supply chains, embedded knowledge, all of which have a value, which somehow also needs to be calculated.

If we were to re-run those calculations a few years later, however, with a possible height restriction on timber construction in force, the decision would be even simpler to conclude, with the main comparator now being:

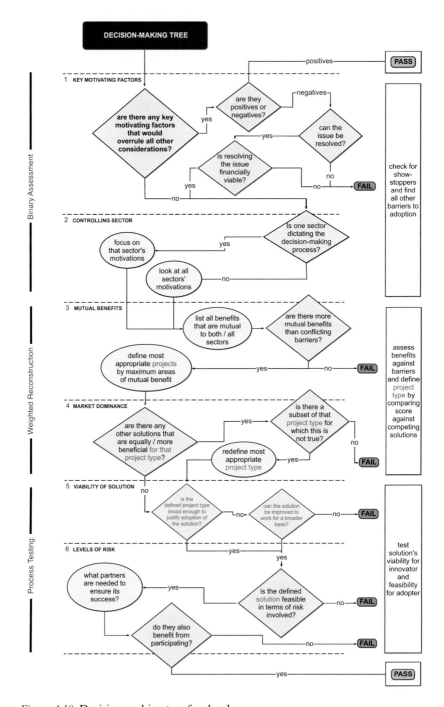

Figure 4.10 Decision-making tree for developers.

The increased development value of X additional stories if built in concrete
vs.
The comparable construction costs of CLT over concrete frame

It is unlikely that this would not now result in a decision to build in concrete without any further debate around the cost of the carbon levy being necessary.

Importantly, this six-stage decision-making tree approach provides a structure for both the promoter and the adopter of an innovation. In the same way it guides the developer's decision whether or not to adopt an innovative solution, it also guides the innovator's decision regarding whether or not a developer or a project represented a market worth approaching to promote their product to. The only difference is in replacing the 'market' being considered for selling in to with the 'product' being considered for adoption. In other words, instead of the innovator deciding whether a market is big enough to justify promoting their product to, the customer must decide whether the product would work for a large enough proportion of their projects to justify the effort involved in adopting it as a solution.

This need for a solution to work 'for all eventualities', however that is defined, is an often overlooked requirement by those promoting an alternative solution. The hold that traditional brick and block construction and strip foundations has over the industry, for instance, is due to its ability to cope with all sites and all conditions for a broad sector of the market. Any competing solution that has to cherry pick the sites for where it is feasible, let alone viable to use, is asking its potential customers to adopt two independent delivery paths, supply chains, skillsets, contracts… The costs involved in that outcome are not hidden, they are just not considered by those expecting to transform an industry without first asking the question 'why?' from the perspective of those who will need to accept their partial solution over a universal one.

4 How Can Those Barriers Be Either Removed or Avoided?

So far we've developed an approach that covers the first three of our four steps:

1   What are the benefits of the innovation?
2   What are the barriers preventing its adoption?
3   Do the benefits outweigh the barriers?
4   How can those barriers be either removed or avoided?

We've looked at and calculated the benefits of adopting a new innovative solution, we've established what the barriers are to doing so, and we've compared the two. Once the perspectives of all those involved have been considered, that comparison will always find there to be barriers that, for some, outweigh the benefits. If that is not the case, the innovation being considered would no longer be an innovation, it would be mainstream practice. The final step is therefore an essential part of the process: how to remove or avoid those remaining barriers.

From the innovator's perspective, the options available for doing this can be categorised into four alternative solutions that cover all eventualities. If something has to change for an innovation to succeed, it can only either be the messaging, something about the product itself, the market being approached or the legislative system influencing the industry's choices. Again, there is a hierarchy to these solutions that defines the order in which these need to be tackled.

i) **Is the message right?** Any solution, innovation, proposition must be known to the market before it can be considered. Dissemination of that information, or ensuring that information is being delivered in the right language for the market being targeted, is therefore the first step, and if that is the only change required, it represents the easiest barrier to be removed.

ii) **Is the market right?** Establishing the right market where there are the strongest synergies with the product being promoted is fundamental to the success of any promotional campaign, especially when deciding where best to launch a product trying to compete against an existing solution. Recognising where there are mutual benefits in a new product succeeding also enables greater collaboration over projects or promotions that cannot be carried out in isolation.

iii) **Is the product right?** A better understanding of the market being addressed and what it is that the market wants can provide a clearer vision of what the product itself needs to deliver, in terms of price point, functionality, regulatory compliance, etc. If the product needs to be changed or improved this will only become apparent after addressing the first two questions.

iv) **Are the policies right?** If there are policies in place that represent barriers to progress for some products whilst benefitting others, these need to be dealt with and will invariably require the assistance of collaborative partners who share similar goals, if not necessarily the same motivations. This is the most onerous option, even before any proposed change in policy is resisted by those benefitting from the status quo, but for that reason often the most necessary.

It's clear from this analysis that this is the innovator's challenge, not the potential adopter's. All potential clients can do is assess the validity of what is being offered, and assess its appropriateness for their needs. Without true collaborative R&D, where the client participates in the design and development of better targeted solutions, all the client can do is seek out the best-fit innovations from a field of contenders. If the messaging is wrong, or aimed at the wrong market, or if the solution just fails to answer the industry's needs, or even just fails to get recognised by a government's behind-the-curve compliance requirements, that approach can easily fail, and there is little else those industry clients can do without being more involved in the design process at the outset.

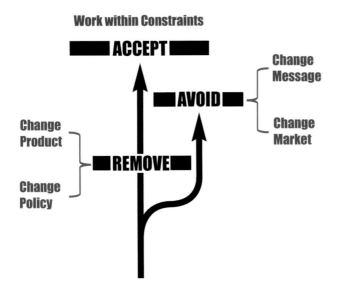

Figure 4.11 The innovator's options.

Conclusion – the Decision-Making Tool in Practice

The first four chapters of this book have tried to explain the need for a better decision-making process, and the way in which this has to capture the benefits of innovations but also the barriers to their adoption, and to do this from a far broader range of perspectives than is currently the case to be truly effective. It has then explained how such a process could be based on a Systems Thinking approach and shown, in principle, how that could work for the construction industry. But what would the tools being proposed look like in practice?

The solutions we need are not here in a book, as they need to exist digitally, online, and structured in the form of relational databases. They also need to be owned at an industry level, made accessible through multiple entry points relating to all sectors of the industry, and maintained through some level of subscription. None of these are insurmountable barriers to such a range of tools being launched, but they are the barriers to be overcome. Were these decision-making tools in place today, they would undoubtedly be of benefit to individuals, businesses, the industry as a whole and therefore the wider economy, but where does the collective will come from to make that transition to a better decision-making process happen in reality?

As was bravely stated earlier, if the approach being proposed cannot be used to make these tools a reality, it is unlikely that they would ever be able to make any other industry innovation a reality either. So that is the challenge. There are two ways that innovations can succeed. They can either grow from their adoption at a grass-roots level and prove themselves organically by their own merit, or they can be promoted / incentivised / dictated from above. If an innovation is right – and that requires not just the right product, but the right pricing and the right timing, then both of these approaches will play a role in that success, and that is what must be aimed at here too.

So, on to Part II and the proof for how transformative this approach could prove to be, with examples of how it has already provided answers to some of the industry's most intractable and divisive problems.

Bibliography

Chartered Institute of Building. (2012). Hackney clarifies: Wood first equal. *Construction Management*, May. https://constructionmanagement.co.uk/hackney-clarifies-wood-first-equal/

Fire Protection Association. (2020). *More details on cladding loans plan*, December. www.thefpa.co.uk/news/more-details-on-cladding-loans-plan

Institute of Chartered Foresters. (2012). *Hackney puts wood first.* www.charteredforest ers.org/hackney-council-puts-wood-first

Lane, T. (2010). Timber frame buildings more susceptible to fire damage. *Building*, August. www.building.co.uk/timber-frame-buildings-more-susceptible-to-fire-dam age/5004548.article

Rittel, H. W. J., & Webber, M. M. (1973). Dilemmas in a general theory of planning. *Policy Sciences*, *4*(2), 155–169. https://doi.org/10.1007/BF01405730

Snow, T., & Greenspoon, A. (2020). Public servants are tired of change-washing – not change: Beware of anyone who claims to have a quick fix for government. *Apolitical.* https://apolitical.co/solution-articles/en/public-servants-are-tired-of-change-wash ing-not-change

Part II

What 'Approaching It Differently' Could Achieve

The following chapters now apply a Systems Thinking approach to the hierarchy of known issues, their causes and their consequences. The purpose of this is not to promote any particular solution, but to show how the solutions reached through adopting a more holistic approach differ from those currently being prescribed. In that respect this is a plea for innovation to be driven by an approach that focuses on asking the right questions, as opposed to dictating the 'right' answers.

The conclusions reached, therefore, are here for discussion and further research only. They all result from approaching many of the same problems already highlighted, not from a different perspective, but from all perspectives working together, and then using this additional insight to suggest alternative approaches, many of which have not been previously considered as mainstream solutions. It seems unlikely that this will represent four full chapters of original thought on matters that have been so comprehensively investigated by so many for so long. A more likely but perhaps more worrying conclusion is that these are all alternative solutions that have been voiced elsewhere, but have not as yet seen the light of day because of that 'barrier at the end of the road' that so successfully maintains the status quo.

To be clear, there are many 'status quos' some of which are relatively new on the scene, which may sound perverse, but helps explain the dynamic that justifies a question-based approach to problem-solving being adopted. The reason why the solutions being discussed in this section are secondary to the discussion around the process used in reaching them is that these solutions are often ephemeral and will need to be constantly re-evaluated based on different locations, clients, sites and the latest technical developments both within the industry and further afield. The pressure to make 'what worked before work again' comes from many angles, but whilst this is a perfectly legitimate place to start the design process, it should never be more than that. Change is necessary, and incremental change, the foundation of vernacular architecture, is the tried and tested way of improving on what is known to work. But the model is easily broken when the pace of change being demanded of an industry outstrips its ability to learn from its mistakes. This is the dilemma that has resulted in government pushing for increasingly radical changes and an industry becoming increasingly resistant to the risks those changes represent. And the longer

DOI: 10.1201/9781003332930-7

this stalemate continues without that incremental, evolutionary development taking place, the greater the risks become, and the stronger the desire to cling on to what is known to be safe. This is not only true of our construction industry but also of our funding and our warranty industries. We have to break this stalemate and that will only happen through the whole industry understanding itself better and finding genuinely beneficial reasons for working together more collaboratively, which is what this next section is trying to bring about.

The Broader Issue of Engagement

The exchange of ideas within the construction industry has always been lumpy. When innovations are eventually proven to be beneficial, that benefit does not automatically get transmitted to others, due to the fragmented nature of the supply chains, strong competition and the adversarial nature of contractual agreements. The architect, once the 'overseer' of operations is now more likely to be just another link in the chain as information is passed from one stake-holder to the next with limited opportunity for iterative design processes or true collaboration to take hold.

Increasingly the industry is dividing into micro-SME operations and large national or international corporations with less and less in between. There are nearly 400,000 SMEs in the UK construction industry now, with SMEs responsible for over 50% of the UK's total revenue, and around 70% of the UK's SMEs being classed as micro-SMEs, employing less than ten people (Mallon, 2019). It is this disparate but dominant sector that has to become more engaged with the many initiatives being introduced before they will have any degree of success in changing the industry as a whole. The tendency, how-ever, is to focus our attention on the progress made by those larger businesses, mostly working within the non-residential sector, where evidence of innov-ation is easier to find, but not representative of the challenges that we need to confront. In fact, housing-related construction, even before including repair, maintenance and refurbishment, as a sector represents around 40% of the total construction market (Kenneth, 2022), but here, innovation, R&D and invest-ment are low priorities (Gerrard & Coulter, 2022). Instead, the black market dominates, with much of this sector dependent on part time, cyclical or migrant workers with an equally dominant Do It Yourself market reflecting the difficul-ties experienced by the public trying to get construction work done at this level.

The many initiatives that have followed on from the Farmer Review, whilst seeming to be increasingly aware of the range and size of businesses that make up the construction industry, have remained predominantly concerned with influencing those fewer, easier to reach and more visible larger firms, rather than the many more smaller SMEs. The result of this government-orientated, top-down, surface level approach has been to further exacerbate the divide between these two sectors, both in terms of the SME companies' likelihood to engage with the initiatives' findings, and the even more important ability for them to influence how these initiatives are being developed at the outset.

The outcome of this has been, time and time again, lower than expected uptake of the changes being prescribed due to the lack of penetration achieved into this sector. But whilst successive papers have built off and honed the ideas put forward by previous initiatives, none have recognised or attempted to tackle this ongoing failure to address the underlying problem of limited engagement. There have been many government initiatives over the years, not all of which need to be referenced here to show the extent to which this is true. Whilst each focuses on different aspects of the construction industry's shortcomings, there is a tendency for them all to suggest that 'this time the industry, or the government, will have to act and follow the report's recommendations in full – unlike last time'. But the lack of engagement at the implementation stage is merely a consequence of the lack of all-round engagement at the development stage.

By way of example, Michael Latham in his 1994 white paper, 'Constructing the Team', recognised the confrontational nature of the industry, which was later picked up by Sir John Egan in his 'Rethinking Construction' paper in 1998 where the focus shifted from recognising these problems towards looking for solutions in a bid to help the industry boost its productivity levels. Building off his previous experience in the automotive industry, his proposal was to interpret the manufacturing industry's approach to business for the construction industry and introduce more efficient working practices. This was not a suggestion that houses should be mass produced like cars on a production line, as the analogy has been more recently interpreted by the volumetric industry, but that the design and planning of that production process should more closely follow that used when designing for a new model of car.

Egan nevertheless did say that 'We have repeatedly heard the claim [from the construction industry?] that construction is different from manufacturing because every product is unique. We [the Construction Task Force] do not agree' (Egan, 1998, p. 18). This in itself is a bold statement – houses, like cars, come in a set number of permutations, but whereas every road surface across the world is effectively the same, every site that is built on is effectively different, and that's a fundamental issue that cannot be ignored. That aside – for now – he then introduced a target-based approach for measuring the industry's progress in terms of percentage reductions in cost, production time, defects and accidents, with the intention of this resulting in a 10% increase in productivity. When these targets failed to bring about any noticeable improvements, Egan returned five years later to look at the skills gap, both technical and professional, the use of training programmes through trial projects and ways to improve the industry's knowledge transfer mechanisms.

The progression here was from recognising the problems being confronted, to suggesting the solutions to be enacted, to helping the industry understand why they should be enacted. The missing information across those ten years of investigation, however, was the industry's perspective on these issues, and ultimately why it was that the solutions being proposed were not being enacted. And this is the recurring theme. Egan himself, instead of persisting with the problem of implementation, then moved on to progressing his own understanding of the industry's objectives by expanding these to include environmental (naturalistic) and societal (therapeutic) benefits, which he again made subject to more market benchmarks and targets (Egan, 2002). That's a lot of sticks, very few carrots, and even less education to help embed yet another round of well-intentioned initiatives.

The measurement of these targets has in itself now become a major barrier to progress and for similar reasons. The collection of the information, needed to both prove that progress is being made and to show that the new approaches being promoted are indeed beneficial, is primarily geared towards meeting the government's own objectives of boosting productivity and complying with their own statutory requirements. For those businesses being asked to collect that data, however, there is no direct or immediate benefit, only an upfront and unmet cost in terms of time and resources. Until the issue of data collection is addressed from the perspective of these individual businesses, this will always be the case, and the data needed to make the case for change will never be sufficient to achieve its goal. This represents one of the many unrecognised concerns preventing these initiatives from being enacted across the construction industry.

This lack of direct benefit to these businesses plus the risk associated with both the development of innovative solutions and with their adoption all goes unrecognised when viewed from the single point perspective of their benefit to the industry as a whole. But none of these issues addressed in isolation, even when resolved to the satisfaction of all the relevant stakeholders, will result in the breakthrough so often expected. There are many of these barriers to adoption, all of which need to be addressed in turn before any meaningful progress can be made. The many white papers referred to here tackle these multiple issues in turn, but ultimately fail to address the one issue common to them all – that of implementation 'at the coalface'. What is it that the construction industry at a granular level needs to get from these initiatives in return for its participation – and to what extent is that being missed because of a failure to engage with the industry's long, hard to reach SME tail during the initiatives' development stages?

The Strategies for Improving Engagement

The intervention strategies that follow next intend to shed some light on this, and are intentionally ordered from the broad to the narrow, starting with how these engagement and implementation problems could be better managed by those instigating the initiatives, together with the consequences of not doing so (Chapter 5). Following on from this, we then look at those interventions that are more realistically within the industry's reach and what can be done from within that remit to improve engagement and adoption (Chapter 6). The final two chapters then look at the tools we need to develop for the industry to be able to solve its own problems (Chapter 7) before showing how they can be used to crack the toughest of all nuts, the housing sector (Chapter 8).

It is the examples in Chapter 5 however – looking at the barriers at the end of the road – that expose the industry's tendency to ignore policy-level problems perceived to be too far removed or just too hard to address, that are fundamental to us making any real progress. We must either accept the limitations that these problems impose on our ability to change the industry and find solutions that work within those constraints, or we must adopt an approach at the outset of 'radical incrementalism', that makes sure every move the industry makes gets it closer to being able to change the political environment it operates within.

In 'Systems Thinking speak', the construction industry is an organism that operates within a political environment, and it is faced with two options. It can either adapt to that environment by its own means, with a degree of assistance from government incentives, or it can try to change that environment to be more beneficial to its own needs, by assisting government in understanding why that option might be more beneficial for all involved. One is a short-term fix and the other a long-term goal, and in reality both objectives need to be addressed simultaneously.

Bibliography

Egan, J. (1998). *Rethinking construction. The report of the construction task force.* http://scholar.google.com/scholar?hl=en&btnG=Search&q=intitle:RETHINKING+CONSTRUCTION#0

Egan, J. (2002). *Rethinking construction accelerating change-a consultation paper by the strategic forum for construction.*

Gerrard, N., & Coulter, M. (2022). Can you survive and thrive as a construction SME? *Construction Management*, 7/11/22.

Kenneth, E. (2022). Construction statistics, Great Britain 2021. *Office for National Statistics*, 000, 2020–2022.

Mallon, S. (2019). SMEs working in the construction sector within the UK. *Planning and Construction News*, 1–3. www.pbctoday.co.uk/news/planning-construction-news/smes-construction-sector/52120/

5 Political Intervention

Changing the Environment That the 'Industry Organism' Functions within

One of the conclusions reached in Part I was that most of the construction industry's 'problems', if interrogated for long enough, resolve themselves into a hierarchy of causes and consequences that lead eventually to a small number of root causes that lie more in the government's than they do in the industry's sphere of influence. It was also established that the construction industry, and the housing sector in particular, is known to spend remarkably little on R&D compared with other industry sectors. Figures 5.1 and 5.2, from Scimagojr. com, the journal ranking website (Scimago, 2018), show how few publications relate to either architecture or sustainability compared with other disciplines, notably medicine.

When it comes to being heard, however, unlike the broader construction industry, which struggles to lobby with a unified voice (Richardson, 2008), the volume housebuilding sector has become remarkably adept at lobbying for its own interests. This represents a heady mix when it comes to driving innovation, as innovation requires R&D, and without that R&D, the lobbying that does take place tends to be focused on protecting vested interests, the status quo and embedded knowledge.

Over the past ten years, a fifth of all Tory party donations have come for the residential property sector, with half of that from just ten individuals (Williams, 2021). So ingrained is the connection between power and land in this country – across all parties – that it is deemed to be outside the window of acceptable discourse, and yet it has been known since Churchill's time and before to be the elephant in the room that dictates our entire economic structure.

> *It is quite true that the land monopoly is not the only monopoly which exists, but it is by far the greatest of monopolies; it is a perpetual monopoly, and it is the mother of all other forms of monopoly.*
>
> (Churchill, 2014)

Calls for land reform have been made in Parliament for over 100 years by prominent but often lone voices drowned out by the sound of self-interest from both Houses. This is the still controversial but stark reality that allows

DOI: 10.1201/9781003332930-8

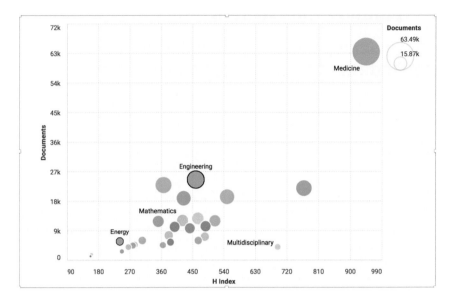

Figure 5.1 Scimagojr bubble chart of research areas.

Source: Scimago, 2018.

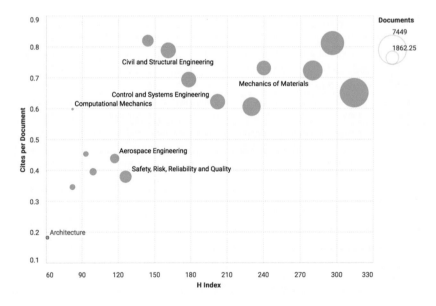

Figure 5.2 Scimagojr bubble chart breakdown for engineering.

Source: Scimago, 2018.

over £10bn to be lost in windfall profits every year to landowners who have to do little more than just be in the right place at the right time, as their land gets lifted from agricultural to developable value (Aubrey, 2018). How we might address this is discussed in detail in the final chapter, because since the 1960s our property market, and as a consequence over 50% of our country's net worth, has become a land-based speculative venture from which it is now almost impossible for us to escape (Murphy, 2018).

This is relevant to innovation, especially within the residential sector, because the price paid for land is dictated by the build cost of a property and its saleable value. This is known as the 'residual land value' calculation, and is tackled in the final chapter. It means that any benefit in developing more efficient ways of building that reduce our overall build costs gets balanced out by the higher land costs that can then get charged by landowners, who become the main beneficiaries from these savings. Whilst property remains a safe bet for investors, this will always be the case, and the financial reward for innovation will to a large extent be lost from the loop.

Rethinking the way we tax our land has had a long and as yet unsuccessful history, which is why, whilst it is not impossible for this to be resolved at some point, it is not something that can be relied upon, but neither can it be ignored as being outside the equation. Should we reach a point when over half the UK's voting public no longer own property, a reality that we are fast approaching despite the government's best efforts to postpone the inevitable with Help to Buy and many other incentive schemes, there may well be a change from policies wholly predicated on that ownership model to ones that recognise the need to tax wealth alongside income.

As will become clear, an awareness of this will colour every decision made from this point on, so fundamental is it to how the construction industry currently operates. In the meantime, however, by adopting a policy of radical incrementalism, there are many less controversial, shorter-term policies that could be lobbied for that would still be beneficial in their own right, whilst also working towards that more radical 'endgame'. With that in mind, the following four examples show how adopting a Systems Thinking approach could lead to some fundamentally different solutions for the construction industry to those currently being promoted by government. They all fall into the category of *'Removing the Barrier by Changing the Policy'*.

The Problem with the Box-Ticking Route to Compliance

Perhaps the least recognised barrier to the adoption of innovative solutions to building in the construction industry is its box-ticking route to compliance. This approach has arguably come about through a need to deal with the ever increasing complexities in how we build, the ever increasing levels of regulation that it has been necessary to introduce, and the ever decreasing time available to make important decisions. The problem with the box-ticking approach, however, is the distance this puts between the problems being dealt with through

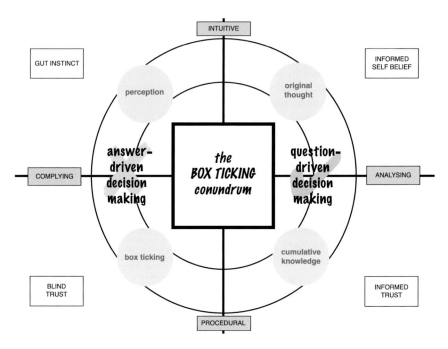

Figure 5.3 The box-ticking conundrum.

compliance and a true understanding of those problems, and why it is that the certified solutions on offer have been deemed to be the best available.

From the perspective of driving innovation, this is detrimental on two levels. Abdicating responsibility for taking a decision about what solution is most appropriate, and handing that over to whatever body provided the choice of predefined options, generates a culture of acceptance based on a superficial level of knowledge. The Grenfell Tower tragedy exemplified the dangers in this approach and even more so the dangers in then removing the regulations behind the predefined solutions to simplify the build process once that knowledge has been eroded. But the box-ticking approach also stifles innovation by not rewarding new thinking. If an alternative approach is proposed, based on, for instance, a societal advantage, for which there is as yet no box to tick, that advantage will not be recognised as advantageous for the decision maker. At best this creates a lag in the system, but more often it represents an insurmountable barrier for innovators to overcome.

Solving this conundrum requires a measure of 'pragmatic idealism'. Ideally we would all have the time to research all options from first principles and reach our own well-considered alternative solutions, and innovation would progress unhindered. In reality, those time constraints are a fact of life in a world where competing means saving on time as much as it does saving on cost. We need processes that 'simplify the complex' and allow us to reach safe conclusions faster, and a predefined set of trusted answers does just that. The suggestion here, however, is that a predefined set of trusted questions would be a more appropriate solution to the issue of time poverty. Ensuring that all the right questions have been asked and considered is a big step towards ensuring a good outcome, whilst achieving two other critical objectives – educating the decision maker about the problem being addressed and allowing the decision maker to reach their own conclusions about what constitutes the best possible solution. The answers to these complex problems will change over time and should be allowed to do so. The questions that need to be asked, on the other hand, have a far longer shelf life.

As a theoretical proposition, that sounds eminently reasonable, but what becomes apparent here and in all of these proposals is that the real barrier to innovative ideas and solutions is rarely a barrier of disagreement with the concept but more likely a barrier of transitioning to the concept. This is particularly true when, as is usually the case, the transition is not a 'flip' but has to include a period during which both options are in operation. Very quickly we're into the realms of long-term benefits as opposed to immediate benefits, incentives to help the innovation through that transitionary period, and with that, the need for backing from a stakeholder who can both see that benefit and finance its adoption. Who fulfils that role depends on how that longer-term benefit can be realised. If the benefit can be easily financialised it could be through private investment, but if the gains are broader and more societal, as is the case with sustainability, we're back to depending upon government intervention.

Applying the decision-tree stages outlined in Part I, if the government is to play the role of key decision maker, what is the message that would be most likely to motivate them into enacting such a fundamental change in how the industry is regulated? What would give them the confidence to stop dictating a direction of travel and instead encourage a question-based approach to problem solving? For government, the question this would raise would be 'how long could such an approach take before showing any positive results, and just how wrong could it go in the meantime?'

Put like that, achieving such a fundamental shift seems a high-risk strategy for any government, and therefore unlikely to be considered, but that doesn't mean the principle cannot be enacted in some other way, and that's the thought process that has to be employed to get results, rather than getting drawn into unwinnable battles. The broader benefits of a question-based approach to compliance are brought back into play later as a pivotal concept, but for now the next question to ask is 'are there any areas within our compliance-based system where current outcomes are particularly unfavourable to any of the government's objectives, and where they might be more open to engaging with an alternative approach?'

Achieving carbon neutrality by 2050 is a statutory obligation, and as a country we're not on target (UK Parliament, 2022). As an industry, we've barely got started, despite a slew of aspirational target-setting initiatives. There are many reasons for why this is proving so difficult, some of which are dealt with in later sections, but for the housing sector, a major barrier to carbon-reducing solutions being adopted is our box-ticking system of compliance, used to measure, amongst other things, thermal performance by means of the Standard Assessment Procedure (SAP). SAP has been in existence since 1995 and was updated every three or four years until the 2012 version that has been in use ever since. The delays that have regularly dogged new releases are a problem, especially when in 2022 the new Part L and O Building Regulations came into force before the updated SAP software was available to allow developers to plan for compliance, but the real problems are due to the methodologies used in SAP being constantly behind the curve in terms of new technologies on the market, and being increasingly incapable of providing an accurate model for low-energy buildings (Murphy et al., 2011). Dealing with the shortcomings of SAP is a priority for government as much as it is for the housing industry having to comply with its often nonsensical requirements, but these shortcomings are also a consequence of a system that relies upon predefined answers to constantly changing problems. The case for a longer-term transition to a more informed, question-based approach to problem solving could therefore be built off this, but in the meantime, we could certainly improve on the answer-based boxes we are being asked to tick, and this is therefore the focus of the next section.

The Problem with SAP

The way we measure the energy efficiency of our housing is known to be flawed on many levels and has over the years resulted in some perverse decisions being taken to gain the system, and little progress made in improving actual outcomes (Better Buildings Partnership, 2019). SAP was first introduced over 20 years ago as a tool for approximating energy use, and was never intended to deal with the demands of a net zero agenda. The tools used for non-residential developments have always been more advanced, and use dynamic rather than static modelling (Elmhurst Energy Consultancy, 2023), but have been deemed too complex for use on individual homes. When trying to achieve results that push the boundaries and go beyond normal compliance, however, dynamic modelling becomes an essential methodology for examining the true performance of different fabric approaches.

In response to the limitations imposed by SAP on those sectors of the industry wanting to be more innovative, alternative assessment procedures began to proliferate that would recognise higher levels of attainment, with PassivHaus certification being the most high profile. Whilst this allowed the highest-performing innovative solutions to be measured more accurately, for the mainstream industry this represented a revolutionary departure from their norms rather than the evolutionary path that they had signed up to. What was needed was for SAP to evolve alongside the demands being made of the industry to make that transition as manageable and risk free as possible – but it has failed to deliver.

The demands for SAP to evolve and embrace dynamic modelling procedures continue to be made, but as in the last section, the government is the key decision maker and, for them, or to be more precise, the powerful housing lobby group, the pros do not outweigh the cons. Dynamic modelling requires calculations to be carried out for each property, as orientation becomes a critical factor, and whilst this is feasible for large developments, it becomes very onerous and costly for individual houses, especially when the process then concludes that each property needs different solutions to be applied. The volume housebuilders' pattern book approach, on which they rely to maintain their profit margins, does not allow for such levels of variation.

This leaves those within the industry demanding change with two choices. They can either continue to fight for the improvements in SAP that will better reflect the benefits that their alternative solutions would offer, or they can accept that those changes are not in the interests of those with most influence over the decision, and look for an alternative solution that will meet with less resistance.

Finding that solution means focusing on what would be acceptable to those resisting any change to the system, primarily the volume housebuilders, whilst still improving on the current situation. Other factors to consider are the cost of implementing a new solution and the simplicity of that solution, in order to minimise resistance to its adoption, and of course the degree to which it would actually reduce carbon emissions, that being the fundamental driver that will engage government.

Taking each of those in turn, the resistance to a dynamic solution comes partly from the cost of modelling individual houses, something that would not be seen as such a problem when designing for blocks of housing. BREEAM, the dynamic modelling solution used by all non-residential sectors of the industry, and already familiar to developers, could however be implemented relatively easily for this substantial sector of our housing market with a simple change in sector definition. Currently, BREEAM Multi-Residential is promoted for use on multi-occupancy residential buildings but its use is not obligatory in many cases.

As we move towards developing more sustainable housing solutions, something that we now need to do at scale, apartment living will become an increasingly common solution, and one that the innovative sector demanding a more accurate assessment procedure will most likely be focused upon. It would therefore seem to be in the government's interests to mandate the use BREEAM in this sector, as that would encourage better sustainable solutions with minimal disruption and without damaging relations with the volume housebuilders on whom it depends in the short term for maintaining at least a base level of housing provision. This is posed as a question to be asked, not an uncontested solution, but getting a response from the Building Research Establishment (BRE) on the matter has so far proved elusive.

Driving the sustainability agenda is a major challenge for the government and one that is met with resistance from many quarters. Any proposition that helps achieve the goals set out under net zero without creating a backlash should be encouraged. The government's first line of defence in ensuring these sustainability targets are met is our local planning authorities, but that defence is crumbling under the weight of an ever increasing workload (Kenyon, 2022). How they could better meet that challenge is the next question asked.

The Problem with Planning

With planning authorities across the UK already struggling to manage their workload with limited resources, any additional legislation, such as that relating to carbon neutrality, only represents a further demand that cannot be adequately met. Developers, keen to limit costs and maximise profits, often push through schemes that pay lip service to these new regulations and provide calculations to prove compliance that planning authorities must then either accept or challenge. Once a submission has been made and is in the system, the clock is ticking and a decision must be provided within the nationally prescribed timescales. In addition, any refused application can legitimately be taken to appeal, resulting in further costs and time spent.

Those are the real-world dilemmas that lead to planning authorities' decision making being skewed by the likelihood of a prolonged battle with a well-financed adversary rather than just being able to focus on a scheme's merits. In the white paper 'Fixing Our Broken Housing Market' (Department for Communities and Local Government (DCLG), 2017), following on from the Farmer Review, the solution was to invest more money in local authorities to better finance their planning departments in return for them speeding up their processes. The proposals also included a promise to simplify the planning process for developers in return for them promising to build more and to build out sites promptly after receiving approvals. As already discussed, these were measures driven by the need to get more homes built, with little attention paid to the unintended consequences in terms of the quality or sustainability of those homes. So whilst that paper did attempt to involve multiple perspectives and give everyone a task to undertake plus a 'reward' for doing so, it was still only considering the outcomes that were relevant within its own remit. In hindsight, a more holistic solution was clearly needed that dealt with *all* the issues from all perspectives.

Granting more funding, resources and authority will always be welcome, but what can be offered is finite and rarely enough to be a solution in its own right. To be viable, the support offered needs to be better targeted, and that requires a far deeper understanding of where the barriers are that need to be addressed, which again starts with establishing who is controlling the decision making in this battle to deliver more sustainable developments. In an industry controlled by regulatory compliance, that ought to be those controlling the regulations, but in a free market economy it isn't that straightforward. First the regulations have to be passed into law, and that requires the acquiescence of those sectors of the industry who lobby government to protect their own business interests in return for their own compliance. Once in place, however, those regulations still have to be enforced, and it is here that the control has shifted from the regulators to the developers.

A recent Climate Change Committee (CCC) annual review of the government's net zero achievements to date refers to the lack of progress in the construction sector and specifically to the '*shocking gap in policy for better-insulated homes*' (CCC, 2022). This dates back to the scrapping of the Code for Sustainable Homes' net zero targets in 2016 and the one million plus homes that have been built since then that will now need to be retrofitted at the cost of the occupiers rather than the development companies. Across the whole industry, though, developers have been able to limit the extent to which they have had to comply with the intent of these new regulations if not also their actual requirements. This has been due to a combination of issues:

• poorly defined targets within the legislation;
• delayed releases of the assessment procedures needed to calculate performance;
• a lack of clarity over the metrics for measuring compliance;
• the inability of planning authorities to check the veracity of the claims made; and
• the threat of developers taking their projects to regions with less combative authorities.

The CCC is an independent body and can only comment and advise. Government has to decide, or in this case regain control of the decision-making process that it has now effectively lost. Their motivation for doing so is clear, as is the developers' motivation for, to be generous, treating change with caution. In a competitive market, even if there is a desire to act more altruistically, unless everyone else is compelled to act in the same manner, profits will fall, at least in the short term, and shareholders will complain. In a similar vein, many local authorities might want to be more aggressive in carrying out their duties, but the need to attract investment means they cannot stray too far out of line with the approach taken by other competing authorities.

Not unusually, therefore, the root cause to be tackled sits within the court of national government, not only in how the regulations and assessment procedures are written and deployed, but in how the local planning departments are supported. This is not necessarily a case of providing more financial support, as stated earlier, but of recognising where the barriers are from the planning authorities' perspective and at the same time finding ways around the government's own barrier of limited available finance.

The fee charged by local authorities for submitting a planning application is fixed at a national level, whereas any developer wanting advice or comments beforehand can submit a pre-application request, or pre-app, which is costed and dealt with locally. This is the point at which the authorities have some degree of control over the system. Unlike when the full application has been submitted, and the fee has been paid and the clock has started ticking, at the pre-app stage, the authority can dictate how the process operates and what it will cost. Crucially, it would also require the power to dictate that a pre-app becomes a mandatory prerequisite for a full application.

This would however solve a number of problems for planning authorities. It would provide the option of a direct income for the planning authority from developers, proportionate to the size and complexity of the development, plus it would also allow them to dictate what is required from them before a full application can be submitted. In return for this the authority could guarantee a smooth passage through planning, at least in terms of its sustainability credentials, without fear of costly delays. The proposal is that this pre-app stage would be used primarily to ensure that the sustainability reports submitted were genuinely achievable, something that can take far more time and resources to establish than the current 8- to 13-week planning window allows for. The onus would now be on the developer to provide enough information to speed up the pre-app process and avoid an extensive counter-interrogation. The option this would also allow for would be for the planning authority to fund the assessment of the sustainability report being undertaken by a qualified third-party business, who could make a far more informed decision without the need for further local authority investment.

The pre-app fee would have to increase considerably to finance this, but from a very low starting position. Developers' profit margins can hinge upon the speed at which permissions are granted (UCL & LSE, 2018), and paying to remove that uncertainty could be seen as money well spent. All that might be needed therefore is for central government to enforce the requirement for a compulsory pre-app 'gatekeeper' stage, to ensure that its net zero requirements are being properly met before an application gets considered on the many other merits that planners are qualified to pass comment on.

The Problem with Funding

This final example represents a systemic problem that sits even more within the government's domain, and focuses on how the construction industry funds its research and development. This is actually a problem for all industry sectors, but construction and the supposedly innovative architectural profession have a particularly worrying track record when it comes to carrying out research that constitutes more than just a site visit. Figures 5.1 and 5.2 at the beginning of this chapter showed where the engineering sector sits amongst its peers in this respect, and how insignificant the architectural profession's research is within this.

Reference has also been made to how the cyclical nature of the industry, closely linked as it is to a fluctuating wider economy, has resulted in the industry's 'survivalist shape' where labour becomes an expendable commodity and little value is put on training or long-term investment. This acceptance of the uncertainty and the lack of continuity that defines the environment the industry must operate within has had a long-term impact, but there is another negative factor that is influencing R&D and ultimately the adoption of innovative solutions, which is its funding mechanisms.

How the construction industry as a whole funds R&D has become an even more relevant factor as a result of the government's latest initiative to transform the way the construction industry operates. This initiative, the Platform Design Approach, is discussed in detail in Chapter 6, but involves the promotion of a 'kit of parts' approach to construction where the industry's supply chain companies work collaboratively to develop fewer, better designed components that construction companies and developers can then use to construct their buildings more efficiently and productively. Whilst this makes sense from a government perspective, keen to standardise the build process without being accused of dumbing down the design of the buildings themselves, it makes less sense to the sub-assembly companies being asked to 'collaboratively' share sensitive information with their direct competition so that they can standardise their products and neutralise their unique selling points.

This is another example of a government-led initiative where a failure to communicate with all the stakeholders or try to understand the consequences of their proposals from alternative perspectives, in this case the supply chain companies, has resulted in a lack of engagement where it is most needed. If there had been full engagement with the supply chain companies from the outset, not just at the point at which they were called upon to deliver, they would have realised that expecting them to design their products collectively for the greater good of the industry was a tall order within such a competitive and adversarial environment. Without seeking out those opinions, the legitimate reasons that sit behind the lack of engagement that has ensued can easily be overlooked, and with them the root causes where the solutions to these complex problems can so often be found.

Collaborative working is now seen as central to reinventing the construction industry and improving on its low productivity levels (Rahman et al., 2014), but little thought has been given to how that collaborative mindset is to be encouraged. It is certainly not a natural state for the industry, and whilst it can on occasion be beaten into an excited state of collaboration through funding and other incentives, when the beating stops, there is a tendency for the collaborating partners to collapse back down into their familiar siloed ways of working. The reasons for this run deeper than comfort zones, and have more to do with the way in which we fund our industry's R&D, since R&D, an industry in its own right, is wholly funded by the capturing of intellectual property rights (IP). If developing a new product requires R&D, and if that R&D requires funding which in turn requires there to be IP to be captured, the incentive is there for the industry to build uniqueness into every new solution, resulting in multiple solutions that effectively perform similar tasks in a slightly different manner. This is clearly the antithesis of what a 'kit of parts' approach is trying to develop, and why, if we seriously want to standardise our supply chain product range, we first need to restructure how we fund our R&D model.

The alternative and more appropriate approach might be to develop these collaborative R&D programmes around an opensource funding model, as has been proven to be very effective in the programming sector (There et al., 2018). In that way, shared resources could be encouraged, and would break down the barriers between companies who could together develop the fewer, better solutions being asked of them, and become part of a collective supply chain with greater security for both themselves and the markets they would be supplying in to. The analogy would be a smaller but guaranteed slice of a bigger pie as opposed to having to fight for a bigger slice of a smaller, more risky pie. It would be equally appropriate to apply that analogy to the IP industry itself, where the strongest resistance to such a proposal would no doubt be felt. Recognising that their perspective on this also has to be considered, and that many of these perspectives will be heavily biased by self-interest, is all part of the same process.

It is nevertheless true that expecting a similar collaborative 'organism' to emerge as exists in the programming world, without first dealing with the adversarial environment within which our supply chain companies currently operate, would be to misunderstand how the industry functions at a fundamental level. In fact, the funding mechanisms are only part of that structure that would need to change. Contractual agreements, routes to procurement and how we define 'value' are all hurdles that the collaborative ideal struggles to clear. Whilst it may be encouraging to know that all these are areas currently being investigated, it is also true that the solutions being proposed are, despite good intentions, still being developed from the single-point perspectives of those proposing them within the construction industry. The following chapter now moves into that realm to look at some of these industry initiatives and how they too could be better constructed if approached from a more broadly informed base.

Bibliography

Aubrey, T. (2018). *Gathering the windfall.* Centre for Progressive Policy, September. www.progressive-policy.net/publications/gathering-the-windfall-how-changing-land-law-can-unlock-englands-housing-supply-potential

Better Buildings Partnership. (2019). *Design for performance: A new approach to delivering energy efficient offices in the UK*, June. www.betterbuildingspartnership.co.uk/design-performance-new-approach-delivering-energy-efficient-offices-uk

Churchill, W. (2014). Land and Income Taxes in the Budget: Edinburgh, July 17, 1909. *Liberalism and the Social Problem*, 15132. https://doi.org/10.5040/9781472581297.ch-017

Climate Change Committee (CCC). (2022). *Current programmes will not deliver Net Zero.* www.theccc.org.uk/2022/06/29/current-programmes-will-not-deliver-net-zero/

Department for Communities and Local Government (DCLG). (2017). *Fixing our broken housing market*, February. www.gov.uk/government/publications/fixing-our-broken-housing-market

Elmhurst Energy Consultancy. (2023). *What's the difference between Simplified & Dynamic Thermal Modelling?* www.elmhurstenergyconsultancy.co.uk/2022/11/03/whats-the-difference-between-simplified-dynamic-thermal-modelling

Kenyon, M. (2022). Nearly 9 in 10 authorities struggle with planning enforcement backlog. *Local Government Chronicle*, November.

Murphy, G., Khalid, Y., & Counsell, J. (2011). A simplified dynamic systems approach for the energy rating of dwellings. In *Proceedings of Building Simulation 2011*, 12th Conference of International Building Performance Simulation Association, Sydney, 14–16 November. www.ibpsa.org/proceedings/BS2011/P_1421.pdf

Murphy, L. (2018). British wealth is concentrated not in property but the land underneath. *New Statesman*, August. www.newstatesman.com/politics/2018/08/british-wealth-concentrated-not-property-land-underneath

Rahman, S. H. A., Endut, I. R., Faisol, N., & Paydar, S. (2014). The importance of collaboration in construction industry from contractors' perspectives. *Procedia–Social and Behavioral Sciences, 129*, 414–421. https://doi.org/10.1016/j.sbspro.2014.03.695

Richardson, S. (2008). Construction's allies with the CBI – industrial strength lobbying. *Building*, July. www.building.co.uk/focus/constructions-allies-with-the-cbi-industrial-strength-lobbying/3118431.article

Scimago. (2018). *What is Scimagojr for?* www.scimagojr.com/index.php

There, C. A., Raymond, E. S., & Tiemann, M. (2018). Open source is 20: How it changed programming and business forever. *ZDNET*, February.

UCL, & LSE. (2018). Planning risk and development. *Royal Town Planning Institute*, April. www.rtpi.org.uk/research/2018/april/planning-risk-and-development/

UK Parliament. (2022). *UK will miss net zero target without urgent action, warns Lords committee*, March. https://committees.parliament.uk/committee/517/industry-and-regulators-committee/news/161468/uk-will-miss-net-zero-target-without-urgent-action-warns-lords-committee/

Williams, M. (2021). 20% of Tory donations come from property tycoons. *Open Democracy, 20*, July. www.opendemocracy.net/en/dark-money-investigations/20-tory-donations-come-property-tycoons/

6 Industry Intervention

Changing How the 'Industry Organism' Functions within Its Environment

The internal structure of the construction industry is no less complex and adversarial than that of government, but there the similarity ends. In government, the halls of power are well established and the roles openly defined, but the opinions of their incumbents are far more opaque, with the reasoning behind the decisions taken becoming increasingly 'hard to reach' the further up the pole one travels. In industry, however, it is the hierarchies and structures that are hard to fathom, but when asked a simple question, most industrialists will provide a straightforward and honest answer. The problem is, they very rarely get asked.

Changing the mindset of an industry starts with changing individual minds, and that requires an understanding of the ways in which the industry is structured, and how contracts, collaborations and transactions are managed. Not every individual will have different motivators, but there is a point at which the message delivered to one SME company will need to differ from the message to another, either because of the sector they operate in, their position in the supply chain, the type of businesses they supply to, or the size of their operation, which even for an SME could be anything from 500 employees downwards.

If convincing an individual business to adopt change is difficult, influencing an entire supply chain will get exponentially harder, as each connected business would have to jump in unison to see any real benefit in doing so. The industry's longstanding battle with embedding BIM (Building Information Management) technology as standard practice is symbolic of this challenge, and all the more relevant because of it being both the vehicle for collaboration within the industry and to some extent a necessary prerequisite for those collaborative relationships to evolve. Many of the initiatives currently being worked upon, and discussed in detail later, such as the Value Toolkit and the Platform Design Approach, cannot realistically progress without that network being in place to transfer data and ideas back and forth along the supply chain. This is an example of how the inter-relationship of causes and consequences needs to be fully understood before any intervention can be successfully implemented, and in this instance, that process does not stop at the implementation of BIM.

DOI: 10.1201/9781003332930-9

If getting the industry to communicate coherently with itself – a prerequisite for promoting the benefits of BIM – has been so unsuccessful for so long, there needs to be some serious thought given to how this barrier to progress can first be dealt with differently.

This chapter therefore begins by proposing a solution to that challenge, before branching out from that to the latest wave of initiatives in danger of failing for being prematurely introduced without the necessary infrastructures being put in place first to support them.

The Problem with BIM

This is a classic 'Catch-22' conundrum. Building Information Management is the vehicle for communication that the industry's supply chain desperately needs for it to work more efficiently and for it to be able to discuss the implications of design decisions during the design process rather than dealing with their repercussions on-site after the event. But implementing BIM across that fragmented supply chain has proven very difficult without there first being a vehicle for communication in place to promote its benefits. This is also a classic example of how it is often the act of transitioning that is a bigger barrier to change than what the industry is being asked to transition to. Most stakeholders can see the benefits of BIM, but few are prepared to invest the time and effort transitioning to something that will only be of benefit once everyone else has committed to the same undertaking.

It is interesting to briefly look at how BIM was first introduced into the construction industry. Paul Morrell, the government's then chief construction advisor, in his 2015 report 'Collaborating for Change' looked to tackle the inefficiencies within the construction industry due to a lack of collaboration. His report followed on from the Egan Report and the Latham Report before that, neither of which had had the impact intended, but Morrell recognised the need for the big stick. In his opinion, if change was going to be introduced, especially something as integrated as BIM, it needed to happen with conviction, with deadlines, and with consequences for missing them.

The recommendations in the report highlight that change is almost impossible without industry-wide collaboration, cooperation and consensus and the professions and other key institutions can lead the way in ensuring that this collaboration is in the public interest.

(Chris Blythe, Chief Executive CIOB (The Edge, 2015))

It was only by legislating for the use of BIM to be mandatory on all government-led projects by 2016 that Morrell was able to start the shift across to a new and collaborative way of working. The fact that it was known to be of almost immediate financial benefit to do so was not in itself enough to guarantee that this new methodology would be adopted, because that benefit could only be realised if all parties agreed to change, and change at the same time.

This fear of 'jumping too soon' and failing to see any immediate benefit from that investment is what was holding back the adoption of BIM, and why its use had to be enforced at a governmental level.

BIM can therefore be seen as both a metaphor for the barriers to change and at the same time an essential element of the change that is needed across the whole construction industry. BIM only benefits the individuals involved when all those individuals adopt it, because the procurement chain it represents has to be complete before it can have any real beneficial impact. But the very act of demanding a collaborative approach strengthens the construction industry where it is at its weakest – in its inherent fragmentation due to the cumulative and historic impact of high levels of self-employment and consequentially low levels of investment. By using the threat of isolation to force a large sector of the industry into adopting – and benefitting – from BIM, this one act seeded the industry with the co-operative framework that it needed to face the many other changes that it also needed to confront collaboratively.

In theory, that is the argument for the 'big stick' approach, and also for it to be wielded, not necessarily by government, but from outside the industry. Once the adversarial norm has made way for a more collaborative working environment, the need to enforce changes that are for the benefit of all those involved should become less necessary, but collective benefit is still very different from individual benefit and calls for an approach that recognises that. Our misguided expectation that individual companies will make decisions for the greater good of the industry helps explain the inertia that is sometimes misdiagnosed as complacency within the construction industry as a whole. Given that understanding, we should be in a better position to suggest how to progress changes that might be seen as necessary by all those involved, but still prove virtually impossible to enact.

Finding a solution to the BIM implementation conundrum is also a necessary first step in this process because so many of the other changes being proposed are far less universally beneficial. If BIM cannot be sold to the industry, what hope is there of promoting more contentious transitions that only benefit some sectors to the detriment of others, and without the communication vehicle for doing so?

The use of EPDs (Environmental Product Declarations) falls into this definition of changes that are seen as necessary for the greater good of the industry, require the participation of the industry as a whole, but unlike BIM, are of benefit only to certain sectors for reasons of internal competition. The reluctance to enforce EPDs is due to these potential conflicts, and the need for EPDs to be adopted fully before they can be said to be acting fairly for all industries. A partial 'Cradle to Gate' adoption (looking solely at the manufacturing process) as opposed to 'Cradle to Grave' (including all operational carbon and end-of-life processes), whilst easier to police, would, for instance, unfairly place timber in a better light than concrete, with no attention being paid to the growth cycle prior to harvesting, the performance figures in use, or the end-of-life methods of disposal. It is unrealistic to expect government to enforce these protocols upfront and across the board, but it would be misleading and detrimental to its own net zero agenda to implement a gradual, public sector led introduction of the principles in an attempt to recognise the need for whole-life costing, whilst failing to compare the impact of different materials accurately.

Off-Site Manufacturing (OSM) represents an equally complex transition for an emerging industry, where the mainstream housing providers only see a benefit from engaging with it at certain stages of the economic cycle – coming out of recession – when labour is in short supply and demand is growing more rapidly than the traditional builders can support. At all other times the OSM industry is seen as competition to be kept at heel in order to protect its own business model.

The relevance of both these examples however is not in why they represent even harder transitions to achieve than is the case with BIM, but in where they *are* managing to establish themselves despite this. It is the model of the 'New Entrants', businesses who have entered the market from other industry sectors with a vertically integrated business model that controls the whole operation from concept design though to sale or rental of the built properties, that holds the key. In this more dictatorial environment, BIM has been made to work, along with many of the other policy decisions that elsewhere struggle to find the co-operative air to establish themselves. It is this level of control over the whole supply chain that would appear to be needed to make the implementation of technologies like OSM and new processes like EPDs achievable.

The question is, does the New Entrants' model represent the way forwards for the adoption of innovative solutions for the rest of the industry to replicate, or does this mark the beginning of the end for the industry as it is currently structured? And the question that stems from that is, what is the fundamental difference between the two models?

If that difference is purely down to the level of control over the component parts of the process, that is what would need to be introduced into the mainstream industry's business model to enable it to replicate the New Entrants' success. Put another way, is the problem one of authority – or, to be more precise, the lack of it? In times gone by, before architects relinquished their professional standing through their innate inability to price a project, there was some oversight of projects whereby all stages in the design and build process were under some degree of central control. Today, that co-ordinating role is no longer present and neither are the fees to support it, which means, amongst other shortcomings, no one is responsible for orchestrating the promotion of

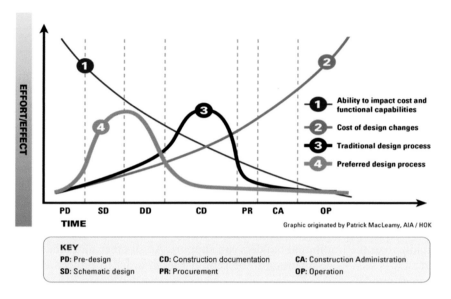

Figure 6.1 The value of early info as delivered by BIM.

Source: Holzer, 2016, p. 470.

BIM. Instead we have devolved the design process into a linear progression of decision making where, in the worst-case scenarios, information is passed forward across incompatible systems from one siloed stakeholder to the next with little awareness of consequences, or any chance of post-rectification before starting on-site.

When the question was asked of the construction industry 'do you agree that architects should be responsible for BIM co-ordination on construction projects?', 80% of respondents said 'yes' (Building Design & Morrell, 2014). Clearly there is a need for BIM to be implemented at scale, and whether this is a vote of confidence or more a passing of the BIM buck, it would appear architects are the natural choice for taking on that task, so perhaps now is the time for the architect to return, not as Master Builder, but as Master Co-ordinator.

To this end, the objectives of the Construction Leadership Council (CLC) in promoting the benefits of smart construction to the UK construction industry are worth noting, as they encapsulate the three examples looked at here (CLC, 2023):

- DIGITAL – delivering better, more certain outcomes by using **BIM**-enabled ways of working
- MANUFACTURING – increasing the proportion of **off-site manufacturing** to improve productivity, quality and safety

- WHOLE LIFE PERFORMANCE – getting more out of new and existing assets through the use of smart technologies.

The following examples only go to strengthen the argument for this approach, as they too are initiatives that will require oversight to bring about their collective adoption.

The Problem with the Platform Design Approach

Platform Design is an initiative that promotes a 'kit of parts' approach to construction where the industry's supply chain companies work collaboratively to develop fewer, better designed components that construction companies and developers can then use to construct their buildings more efficiently and productively. There is a disconnect, however, between what was initially expected from the industry's supply chains under this initiative and how those supply chain companies currently operate, briefly discussed in relation to the R&D funding model in Chapter 5. The collaboration that would be required between companies that currently have to exist in an adversarial environment to collectively develop a reduced range of standardised products cannot be expected to materialise without first addressing how the broader industry operates. A fundamental part of this problem is the way in which R&D funding is predicated on the capturing of IP. This is integral to how the funding industry finances R&D, but in this instance, it is in direct contradiction with what the Platform Design Approach (PDA) programme is trying to implement, in that it encourages the development of multiple solutions to the same problem. But this is only one of a number of reasons why the supply chain companies are not jumping over each other to engage with a programme that is arguably the most practical, viable solution yet to be proposed for the industry to adopt.

Platform Design was one of a number of programmes funded by the £170m Construction Sector Deal allocation from the Industrial Strategy Challenge Fund (ISCF) in 2017. From the outset, the PDA programme was designed to be inclusive of all sectors of the industry to ensure a broad range of perspectives were incorporated into its development. As is usually the case in these programmes, representatives of the industry were invited to be part of the process on a voluntary basis, and in this instance were expected to allocate around 300 hours of their time over a two-year period, and travel to attend meetings, seminars and events on a regular basis. Before too long, however, most of the businesses representing the SME supply chain companies had dropped out of the programme. Little attempt was made to establish why this had happened or deal with the consequences, but that was the point at which the process diverged from its intended objective.

The reason why the predominantly small SME supply chain companies left the process was because, unlike the far larger developers and contractors, they could not afford to maintain a presence and neither could they see any immediate benefit in them doing so. But without their input, the development of the programme became increasingly focused on the far more immediate benefits of the PDA as seen from the perspective of those remaining. For the industry procurers, the purpose of the PDA was to standardise the supply chain's products to simplify their design processes and make their products more affordable to purchase. For the government, the aim was to boost productivity by simplifying and speeding up the whole build programme. These are all legitimate aims, but when presented to the supply chain companies as a fait accompli, their response was muted. Had they been consulted throughout the programme's development, they would have perhaps made the following points:

- This is one of many government-driven initiatives that we have been asked to engage with, and from past experience, we have little faith in it being supported for long enough for it to have any genuine impact.
- As usual, we are being asked to invest time and resources into something at our own risk with no evidence that we will directly benefit from doing so.
- The promise of a larger aggregated market resulting from this approach for us to sell in to needs to be quantified before it can be factored in to our financial decision making.
- The risk associated with being expected to share our IP and sensitive price information so that it can be accessed by our competitors has not been considered.
- The danger for us is that we lose our IP by all manufacturing the same products only to become one of many suppliers vulnerable to being beaten down on price.

Those are five good reasons for not engaging with a process that has been developed primarily to benefit those instigating it, and without the supply chain companies' engagement, there will be no PDA programme. None of these problems however are insurmountable, and it's clearly in everyone's interest to overcome them. This is how a re-run programme with full engagement might have looked:

One of the deliverables requested by the initiative's instigators was a database of supply chain manufacturers so that those procuring could see what was available, where, and in what quantities. A perfectly reasonable request, but at no point was it suggested that the supply chain companies might equally want to know who the procurers were, their requirements and the degree to which they had managed to aggregate their demand. A data source capturing both supply and demand is what is needed to encourage that missing two-way street of information, and this is how it would need to be structured to answer the concerns of those supply chain companies not as yet considered:

Minimise the Burden

Firstly, there needs to be a recognition of the time and effort that goes into providing information to different accreditation bodies, for framework agreements, tenders and procurement processes. Minimising this burden reduces a major barrier to engagement and can be done in a number of ways:

- Not all information is needed at the outset of a negotiation and can be sequenced dependent on the stage of the process reached.
- Assessing the credibility of a supplier can be done by referring to a whole range of existing qualifications or recommendations / testimonials, rather than introducing further bespoke measures of attainment.
- If there is a value to be attached to this information and the time and effort that goes into collecting it, it should be paid for, even if that is only as a down payment on further work.

Recognise the Value of IP

Secondly, there needs to be a recognition of the sensitivity of the data being asked for. Whilst the basic information defining the product and where it is manufactured should be provided openly, it seems quite reasonable that more sensitive information should be made available at the owner's discretion and only once a dialogue has taken place and some level of commitment established.

Prove the Benefit

Thirdly, this needs to be a two-way street of information with equal emphasis put on the market evidence made available to the supply chain companies, regarding both size of market and the market's design requirements.

Guarantee the Support

And finally, there needs to be proof of ownership of the PDA programme at an industry level to give the supply chain the confidence in this not being just another short-lived panacea with no real long-term governmental support.

The proposal put forward using this approach was to separate out the information being requested into three staged categories that related to when that information needs to be known, and then to also factor in the sensitivity of the data at each of these stages with controls in place for the data provider to decide how that can be accessed.

1st Capability: What product is being produced, where, and in what quantity.
  2nd Credibility: Company levels of reliability, quality and professionalism.
    3rd Compatibility: The synergy between a supplier and a procurer's approach to business.

The first of these, Capability, captures the standard information already in the public domain, plus a second tier of more sensitive product price and manufacturing data where the provider can restrict access until a contract is up for discussion.

The second, Credibility, relates to any assurances held, either in the form of existing company and product accreditations, or through testimonials / recommendations from other clients. The second tier in this instance covers any further proof or qualification procedure specifically requested by a client for their project and any more sensitive financial company information required for insurance cover purposes.

The third, Compatibility, relates to another initiative, the Value Toolkit, that was run alongside the PDA programme, looking at how to redefine value in the procurement process to also encompass societal and environmental benefit. Again, there is an initial statement of intent that can easily be made at the outset without having to enter into a more nuanced discussion that can only really take place once a relationship has been agreed and a project introduced. The Value Toolkit and its mutually beneficial relationship with the PDA is discussed in more detail in the next example.

The graphic in Figure 6.2 represents the way in which these criteria can be met whilst also incorporating the other initiatives in development at the time of writing that also need to be incorporated into a more holistic system of delivery to achieve their full potential.

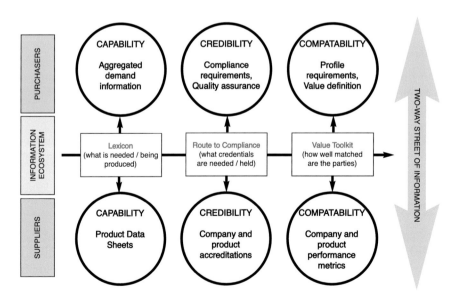

Figure 6.2 Supply chain database model for mutual engagement.

Pulling all these new concepts together is important from a development perspective but also from an implementation perspective. For supply chain companies to be able to engage with yet more procedure, these concepts must be delivered in a way that is easy to access and understand, and most importantly, in a way that makes the benefit to their businesses clear.

As yet, this vehicle does not exist, and without it, the messaging that would help generate some genuine engagement will not penetrate the industry beyond the top-tier developers who directly benefit from government procurement schemes, and, as has already been stated, without that supply chain engagement, there really is no PDA programme.

The Problem with Metrics

As with BIM being a necessary prerequisite for the implementation of the PDA by providing an essential channel of communication down the supply chain, so too are the metrics used to measure performance a necessary prerequisite for the implementation of the Value Toolkit, or any other tool promoting the benefits of an innovative way of working, managing, or building within the construction industry. The recently launched Value Toolkit is central to how the government intends to reduce carbon emissions across the construction industry, by rebalancing its decision-making process around longer-term environmental and societal benefits. So it is imperative that we first resolve our longstanding problem with how we define, measure and capture the information needed to promote more sustainable methods of construction in the form of comparable metrics.

Looking at this conundrum from a Systems Thinking perspective, there can be shown to be no less than six fundamental issues with how these metrics are being captured that need to be addressed: Profitability, Relevance, Immediacy, Security, Comparability and Simplicity. If these issues are not addressed, it is reasonable to assume there will continue to be only limited engagement from the industry, and little worthwhile data will be gathered. This data is the evidence that the industry needs to prove for itself the benefit of transitioning to a new way of working, and without that evidence there will be no willing transition, only enforced compliance.

Industry metrics are currently being seen as a way to promote the benefits of the changes being proposed, but are focused almost entirely on the government's objectives of boosting productivity and meeting its carbon emissions targets. Metrics on their own do not change mindsets, and neither do arguments based on 'the greater good of the economy' or even the industry. Giving these metrics some relevance from the perspectives of those being asked to embrace them means going beyond the promotion of the broader benefits their use may lead to, and dealing with the legitimate concerns that are preventing individual businesses from engaging with them, which are a lack of:

Profitability: Too focused on productivity at an industry level, instead of profitability at a business level

Relevance: Not relevant to how individual companies define value
Immediacy: Only of use collectively at a later date
Security: Require companies to share sensitive price information with their competition
Comparability: Too many variables, alternative systems and definitions to produce any meaningful data
Simplicity: Too complex to understand and time consuming to collect.

We need to remember what the metrics are for. They are meant to increase confidence amongst individual companies that need to be convinced that the changes being promoted will benefit them. The approach to metrics currently being promoted, however, requires:

- everyone to fall in line and agree to work with a universal set of measurement criteria;
- all of industry, including traditional builders, to honestly report their levels of performance, no matter how poor;
- the array of uncontrolled variables involved to somehow be accounted for in the measurements collected; and
- individual companies to act collaboratively for the greater good of the economy.

These all represent fundamental flaws in current thinking which are not being addressed. Progress will only be made when the process of collecting metrics is immediately beneficial to those being asked to collect them. Businesses don't want statistics, especially if those statistics only focus on the positives – they want honest data that helps them make decisions that will benefit them directly. Instead they are being subjected to a constantly changing array of competing systems being promoted by different industry bodies as the definitive metrics needed to promote MMC solutions. Some of these are shown in Figure 6.3 as put forward by the CLC, Buildoffsite, and the Construction Innovation Hub. They are broadly similar, and can be broken down into Cost, Time and Quality related, plus a further category focused on broader Societal and Environmental impacts. What none of them capture, however, but which have been added in here beneath, are the more practical aspects of transitioning, or in other words, the barriers to adoption.

When now looked at from the perspective of those businesses being asked to gather this information, the approach needed would seem to be more of an information-based system that helps companies understand for themselves what the questions are they need to be asking to help them make the right decisions. The data and the statistics will flow from that, but data and statistics provided without context do not in themselves change opinions – that's done through knowledge and understanding linked to what's seen to be relevant.

The solution being proposed forms part of Chapter 7, where the tools we need are discussed in detail, but the answers to these six questions that are not being addressed are, when asked, both simple and achievable:

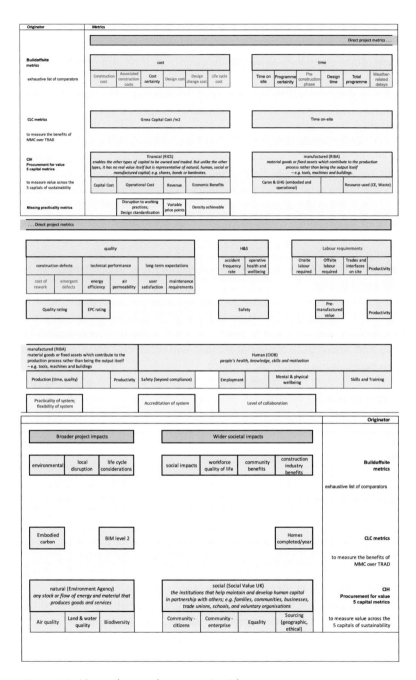

Figure 6.3 Alternative metric systems (not) in use.

Profitability: *Relate the metrics back to recognised cost centres*

Companies need to see how any costs or savings would impact on their bottom-line profits, not on the industry's profitability as a whole.

Relevance: *Allow companies to weight the metrics to their own business model preferences*

Societal and environmental benefits need translating into how they will impact upon individual business models before a company can make a case for their inclusion.

Immediacy: *Make the data being collected immediately relevant to the project in hand*

For data to be worth collecting it must be immediately useable in the decisions being taken now, not for the industry as a whole at some future date.

Security: *Anonymise the data before pooling it into a national data resource*

If data is to be genuinely reported it must be anonymised to allow those businesses not performing well and using traditional methods of construction to engage with the process.

Comparability: *Internalise the comparison system and allow companies to choose their own metrics*

Achieving an industry standard for all metrics is both idealistic and unnecessary. The comparison of value being sought is between two alternative approaches on the same site at the same time, without being skewed by the multiple and uncontrollable variables that will otherwise mask any useful information being derived from the exercise.

Simplicity: *Make sure the system being proposed is simpler than what is currently expected of them*

Individual control over any system being proposed allows for it to be managed in a way that is familiar to that company, and therefore far more likely to be used constructively, rather than being seen as another box-ticking exercise.

The Problem with Retrofit

This final conundrum is in a different league, and from a Systems Thinking perspective also sits in a different space on the concentric circles of organism and environment. In fact it spans them all, as it encompasses the truly existential question of climate change right down to recognising the role of individuals and changing minds at that personal level.

It has been left until last for that reason. Nothing encapsulates the need for a holistic approach more than the retrofit challenge, and with that, the need to look for solutions outside of the constraints of the immediate problem.

The case for investing in buildings retrofit as part of economic recovery remains strong – there are major benefits in terms of emissions reduction, cost savings and wider benefits; it is labour-intensive and spread across the country. Fundamentally this is something that needs to happen on the path to Net Zero and supply chains are well below the levels they need to be at in order to deliver Government commitments on fuel poverty, energy efficiency and heat over the next decade.

(Climate Change Committee (CCC), 2022, p. 166)

In response to this, over 300 local authorities have now declared a climate emergency, but all have picked unrealistic dates by which they intend to make their towns and cities carbon neutral, many as early as 2030. Part of that process will be to retrofit all their properties and reduce their carbon emissions to zero. This will not happen because it is patently impossible, certainly within the timeframes being proposed, but there will no doubt be many misguided and expensive attempts to achieve something, to show willing if nothing else. The concern is that, at best, this will prove to be a distraction that will have no meaningful impact, but more likely result in some expensive and unintended consequences that will then need to be managed in addition to the main task at hand.

In order to meet the urgency of the climate emergency declared by local and national government, there needs to be a step change in the way in which these declarations are treated within their decision-making processes. Without it, there is a risk both that Climate Emergency Declarations simply become worthless political statements but, more importantly, that the opportunity for effective action will be missed.

(Environmental Law Foundation, 2021, p. 2)

There's a very good reason why the Climate Change Committee (CCC) has planned for a transition to net zero over 30 years and not ten. In this sector alone, the challenge of retrofitting more than 26 million homes – that's over a million homes a year – would at today's estimates cost well over £1 trillion. Structural changes of this magnitude take time to put in place, with the correct legislation, technical development, training and behavioural change messaging all essential precursors to the more visible action of 'just doing something'. There are in fact many knowledgeable bodies working in this space, but they are not necessarily being listened to by either government or the businesses on the ground being charged with carrying out these programmes of work. Ultimately, we have created a problem of reliance upon legislation over knowledge, where the impossible task of keeping an entire industry up to date with the latest 'best practice' has taken precedent over providing them with the tools to work out for themselves what that best practice should look like and why it might be in their best interests to engage with it. For that to work, of course, the messaging has to be balanced and trustworthy, which takes us back to how best practice is being defined, and from whose perspective.

As in previous cases, the retrofit programme is being driven primarily by the government's agenda, which is not in itself a cause for concern. It is more the government's dominant perspective on how to promote and implement that agenda that needs to be questioned. In its simplest form, the agenda is to achieve a net zero economy, with that ambition encompassing the need to become energy independent, to level up Britain, to reduce fuel poverty, plus some other more partisan objectives. None of these are necessarily aligned, especially when timescales are introduced into the equation, not least the timelines dictated by our parliamentary cycle. The political influence this introduces into the decision-making process cannot be taken out of the equation, but it must be recognised for what it is, and dealt with accordingly.

The recent decision, for instance, to allocate £700m to the construction industry's retrofitting programme through the SHDF (Social Housing Decarbonisation Fund) to be spent before the 2024 election (Department for Business, Energy & Industrial Strategy, 2022) can be viewed in one of two ways. The CCC has expressed its deep concern with the (slow) speed at which the government has approached the retrofitting of our housing stock, and has said in plain language that we will not meet our zero carbon target if it does not address this one issue in particular. This recent investment in decarbonisation is therefore welcome news, but the implementation strategy will dictate how effective that investment actually is. Wanting that funding to be allocated and spent in its entirety within two years, for reasons that don't need explaining, is not a promising start.

Developing a more desirable strategy starts with gaining a better understanding of all the stakeholders' motivations, including the government's. As with the government's own internal drivers, these will not be aligned and will need to be placed in a hierarchy that recognises both the importance of the issues themselves and the drivers of those controlling them, before establishing a strategy that deals with those conflicting needs as practically and fairly as possible. To test this, we first need to know who the stakeholders are, again taking the housing sector as the most complex example:

Central government, by department
Local government authorities
Developers / builders
Product and solution providers
Warranty companies
Design professions
Building owners
Building occupiers

Establishing a hierarchy of influence across these stakeholders is less straightforward, as there are different answers to this depending on the business model, tenure, procurement routes, etc. What is worth establishing early on, however, is the fact that there is currently no direct profit in retrofitting at scale for the developer or the occupier without funding. This immediately skews the influence in favour of those offering the long-term finance needed, which will either come from government, local or central, or private sector investors. Payback periods are therefore clearly an important factor in this decision-making process, and with that, the need for some complex calculations to be carried out, taking into account the many unknowns that will influence these – government policy, future innovations, fuel prices, climate change. . .

It soon becomes apparent that retrofit needs to be led by a national strategy, and even pronouncements by local authorities are of little relevance without that, as they will be dependent on that national programme of legislation, funding, training and education. So making sure that our national strategy is properly informed becomes critical to the success of any local programme or individual project. But how do we know how informed government decision making really is? Can we be sure the right decisions are being taken for the country as a whole, and that we're not being governed by snap decisions, misinformed by one sector's subjective lobbying, politically or even personally motivated – any of these are possible and quite often the most likely outcomes (Ilott et al., 2016).

Looking specifically at the allocation of £700m to retrofitting through the SHDF, the questions that arise from this are – or should be:

- Politically, does that full amount have to be spent before 2024, or could it just be allocated by this deadline to those successful applicants, to be paid out in phases linked to delivery and performance?
- Is there a figure for the anticipated cost of retrofitting all UK properties and to what level of performance?
- Has enough thought been given to who will be training up the skilled workforce needed to carry out the works being funded?
- Has adequate funding been allocated to that task, and what stage is it at as a necessary precursor of any retrofit rollout programme?
- Will both the legislation and the assessment procedures for measuring performance to the new standards be in place in good time?

The answer to all these questions is arguably 'no', and this is before we start to look at the actual rollout strategies themselves and the solutions on offer, which are far from being ready for scaling up. But none of this means we can't start now, given the right strategic approach.

The retrofit challenge is not limited to just developing the right technical solutions but covers financial, political, societal and environmental considerations that are as important and at least as hard to resolve, not least because they are all interrelated and often in direct conflict. A successful solution needs to satisfy all of these demands, and where they are in conflict, decisions need to be taken about where compromises are to be made, and what the priorities are that those compromises need to be based upon:

Technical: solutions must be certified and installers must be qualified. No solution should be rolled out at scale without solid evidence of its safety and efficacy. All solutions should be seen as part of a package of wider measures and not assessed or carried out in isolation.

Financial: viability is key and should be judged on a solution's payback period in terms of installation costs measured against energy savings and the actual carbon benefit achieved.

Behavioural: a failure to address occupier awareness and understanding of what is being proposed can lead to unintended consequences and a radical reduction in benefit. Buy-in is the essential hurdle to clear that often requires an educational programme to run alongside any intervention.

Environmental: this often entails a compromise between operational and embodied carbon, but should also consider the source of the products and materials being proposed and the degree to which the sustainability of their manufacture and transportation can be verified.

Locational: Ultimately, the feasibility of any solution 'at the coalface' will decide its fate. Many solutions that work well on paper and can be calculated to show an impact run into practical problems on-site that limit their potential to a point at which they may no longer be viable.

Governmental: the government of the day's own political agenda cannot be ignored, even if it runs counter to the logic employed to satisfy all other factors, as the financing of retrofit will be largely dependent on either local or central government funding and dictated by the policies that sit behind it.

Developing a retrofit strategy taking all these factors into account would stand a reasonable chance of being successful, but the bottom line is still the eyewatering cost and how this can be financed. Barring any silver bullet to retrofit being discovered, this will remain a question for government to both ask and answer. The current best-practice cost of a 'deep retrofit' solution is still above £60k per property, with payback periods, both in terms of cost and carbon, regularly extending beyond the life expectancy of the properties being improved (Rodrigues, 2023). Whilst this will be improved upon to some extent through economies of scale, the enormity of the task cannot be overstated, and the practical realities that will stand in the way of any comprehensive roll-out programme must be taken seriously.

At some point this discussion needs to be taken out of its box, and some alternative solutions considered. If the government did want an honest answer to the question about the total cost of retrofitting the UK's housing stock, it could easily be given, and would then allow for a comparison to be made between that figure and the cost of investing in a 100% renewable energy supply for the UK as a whole, something that will have to be addressed at some point in any case. This however remains a substantial 'unknown', especially considering the level of uncertainty that exists around the future cost of fossil fuels on which all payback estimates are based. It is not inconceivable that the differential would be so great as to radically influence the government's whole retrofit strategy, possibly allowing the industry to focus on a far more realistic programme of works based on reducing fuel poverty rather than retrofitting all properties to a future homes standard – bearing in mind that even then there would still be a further 20% of carbon savings needing to be offset.

Currently, this does not feel like the informed position that should be sitting behind such pivotal and far-reaching decisions which will impact us all not only financially, but societally and environmentally for generations to come.

Bibliography

Building Design, & Morrell, P. (2014). Investing in BIM: A guide for architects. *Building Design*. www.bdonline.co.uk/home/the-bim-white-paper/5034325.article

Climate Change Committee (CCC). (2022). *2022 Progress Report to Parliament*, June. www.theccc.org.uk/publication/2022-progress-report-to-parliament/

Construction Leadership Council (CLC). (2023). *Vision and mission.* www.constructionleadershipcouncil.co.uk/mission

Department for Business, Energy & Industrial Strategy. (2022). *£1.5 billion to improve energy efficiency and slash bills.* September. www.gov.uk/government/news/15-billion-to-improve-energy-efficiency-and-slash-bills

Environmental Law Foundation. (2021). *Local urgency on the climate emergency?* October. https://elflaw.org/news/local-urgency-on-the-climate-emergency/

Holzer, D. (2016). BIM's seven deadly sins. *International Journal of Architectural Computing*, 9(4), 464–480. https://doi.org/10.1260/1478-0771.9.4.463

Ilott, O., Randall, J., Bleasdale, A., & Norris, E. (2016). *Making policy stick – tackling long-term challenges in government*, December. www.instituteforgovernment.org.uk/publications/making-policy-stick

The Edge. (2015). *Collaboration for Change, The Edge Commission Report on the Future of Professionalism*. Designing Buildings Wiki. www.designingbuildings.co.uk/wiki/Collaboration_for_Change,_The_Edge_Commission_Report_on_the_Future_of_Professionalism

Webb, S. (2023). How 'deep retrofits' could end up costing you a huge £69,000 – more than double the figure claimed by the government. Homebuilding, January. www.homebuilding.co.uk/news/how-deep-retrofits-could-end-up-costing-you-a-huge-pound69000-more-than-double-the-figure-claimed-by-the-government

7 The Missing Tools

The Tools We Need, How to Build Them, and How to Promote Them

So far we've looked at a number of interventions that have failed to have the impact expected of them, discussed why that might be so, suggested an alternative Systems Thinking based approach, and tried that out to see where it gets us.

The main reason given for these low impact levels has been the 'single-point perspective' that has informed the development of those strategies, meaning that, when implemented, engagement across the industry has been low due to there being little perceived benefit for the stakeholders whose perspectives were not considered at the outset. The proposed alternative has been to adopt a more holistic approach to ensure buy-in at that implementation stage. But buy-in still needs to be bought – it cannot be relied upon just because the research suggests it would be beneficial for all stakeholders to engage with the process. This is why, although Systems Thinking can provide a better methodology for tackling these complex problems, it is still necessary to develop the accessible tools that would enable those hard to reach stakeholders to learn about and engage with these initiatives and reach their own conclusions about the potential benefits they could offer. That still represents a hurdle yet to be cleared.

It was stated earlier that promoting a new way of thinking was no different to promoting a new way of building. Both have to be seen as better – substantially better – than the existing solution, which can be interpreted in many ways: For a new way of thinking about problem solving, the ability to arrive at 'better' strategy decisions ought to, ideally, be the main driver, but being quicker, more convenient or simpler to understand will be seen as equally, if not more, important – even when there is no upfront cost involved in adopting it. New businesses models that focus on convenience rarely seem to fail.

Of the three hurdles, knowledge, motivation and ability, used earlier to define the barriers to be overcome, the approach being proposed here has therefore been developed to satisfy that 'motivation' of 'ease of use' before then dealing with the practicalities that might still make the 'ability' to engage impossible, or too great a risk. That left what might look like the relatively simple hurdle of 'knowledge' – raising awareness of an approach, explaining the concept and how it could be of direct benefit. But saturated as we are in information from

DOI: 10.1201/9781003332930-10

all quarters, this is in fact arguably the hardest and most time-consuming part of the entire operation, and why developing the right tools for achieving this are so important.

The three tools discussed next have evolved out of looking at the strategies currently being developed for the industry, recognising why they might struggle to get the engagement they deserve, and then addressing that gap found at the implementation phase of the process. In that respect they could be seen as facilitating tools that other existing approaches can 'plug in' to. This is both important and intentional, as promoting a solution that in itself helps to promote existing processes is far more likely to be accepted than one that attempts to replace them. Positioning these tools within that promotional role is basically suggesting that all we now need to do is translate the message into a language that is more likely to be heard – one that explains the benefits from the perspectives of those un-consulted stakeholders.

The Metriculator – a Solution to the Metrics Conundrum

> *Manufacturing and construction: Progress in this sector is hard to ascertain, due to the poor availability of relevant data across the various sub-sectors which critically limits monitoring and evaluation of policy implementation.*
> (Climate Change Committee, 2022)

The metrics conundrum has been outlined already, along with the factors not currently being considered and how those factors could be addressed:

Profitability: *Relate the metrics back to recognised cost centres, not just to the industry's overall productivity*

Relevance: *Allow companies to weight the metrics to their business model preferences*

Immediacy: *Make the data being collected immediately relevant to the project in hand*

Security: *Anonymise the data before pooling it into a national data resource*

Comparability: *Internalise the comparison system and allow companies to choose their own metrics and use their own data*

Simplicity: *Make sure the system being proposed is simpler than what is currently available to them*

Metrics are a problem the industry has been grappling with for almost a decade now, but recently the need to resolve this has been escalated by the launch of the Value Toolkit, which is equally dependent on the gathering of data relating to building companies and their projects. Whilst this initiative is based upon collecting much of the same information needed to promote the benefits of modern methods of construction, it is now being gathered to measure the societal and environmental value in a project, as opposed to just bottom-line profit. The clear link between the two is the advantage to be gained from adopting MMC solutions in achieving these broader benefits, and what is exactly the same is the list of barriers above that still need to be overcome for this to be successful.

It was stated earlier that the industry has not necessarily tired of change so much as it has of change management theories, which would suggest it's the messaging rather than the message itself that's the problem. The Value Toolkit recognises this and has been developed to give individual businesses a degree of control over a process that wants to capture the broader benefits of construction to arrive at a fairer procurement process, whilst recognising that those broader benefits are rarely of any immediate concern to an individual business in the same way as bottom-line profit always will be. So far so good, but recognising the problem does not in itself provide the solution. The danger is that, despite these efforts, this could still be seen as yet another hoop to jump through, a box to tick, to show that thought has been given to the societal and environmental impact of a project, whilst still deferring to the lowest-priced provider when making that final decision.

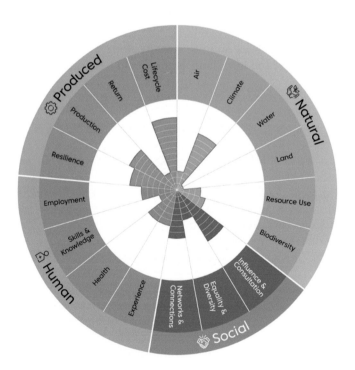

Figure 7.1 The Value Toolkit wheel.

Source: The Construction Innovation Hub, 2021.

The missing link, again due to the lack of focus given to how small businesses have to operate, is in how those broader benefits have not been translated into actual benefits for those individuals who we need to start thinking differently about value, and not just see this as complying with another access-to-tendering requirement.

The solution being proposed here is therefore not an alternative to the Value Toolkit approach, but a necessary first step towards changing mindsets that will make the industry more knowledgeable and willing to engage with the process of data gathering. The basic premise is to start with the known reasons why there is a reluctance to engage with metrics gathering for whatever purpose, and confront them head on. That means being honest about the more practical difficulties around transitioning to the new way of working the metrics are hoping to promote, and not just focusing on the benefits that might flow from doing so. This is the 'what's in it for me?' approach, and for the Value Toolkit initiative, starts with a request that even the most time-challenged business would respond to once the benefit in doing so has been clearly explained.

The Matriculator in Stages

The underlying principle of this tool, based on the six barriers to adoption outlined above, is that it has a direct relevance to businesses in the way in which it allows them to gather their own data and make their own internal comparisons, but for that data to then be aggregated anonymously for the industry as a whole to learn and benefit from. And to make this a less onerous process, the tool has been designed to simplify the interface by keeping the complexity of the calculations being made 'hidden', but accessible to any business that thinks the assumptions on which they are based need refining. Those adjustments then also become part of the data-gathering exercise as they collectively measure the industry's understanding of its own processes.

Step 1 – The first step is to get individual businesses to privately define their own business model by scoring themselves honestly against a range of eight metrics. This has to be a genuine reflection of a business's priorities, so it has to be an internal process. Although these are simply defined concepts and just comparative measures, the process will nevertheless force most companies to consider their own business models in ways they would not normally do.

Step 2 – Then, for the specific job currently being discussed, the company enters a cost breakdown based on their own in-house cost centres, selected from an exhaustive list of possible choices, together with a further simple priority choice for that project.

This is the extent of the information-gathering exercise for the Metriculator to then provide an assessment of how the many metrics in use across the industry actually influence each of those cost centres, to what extent, and how much that would vary with the different methods of construction being considered. The complexity of the process is in the 'behind the scenes' algorithms that use information from the company's business values entered in Step 1 together

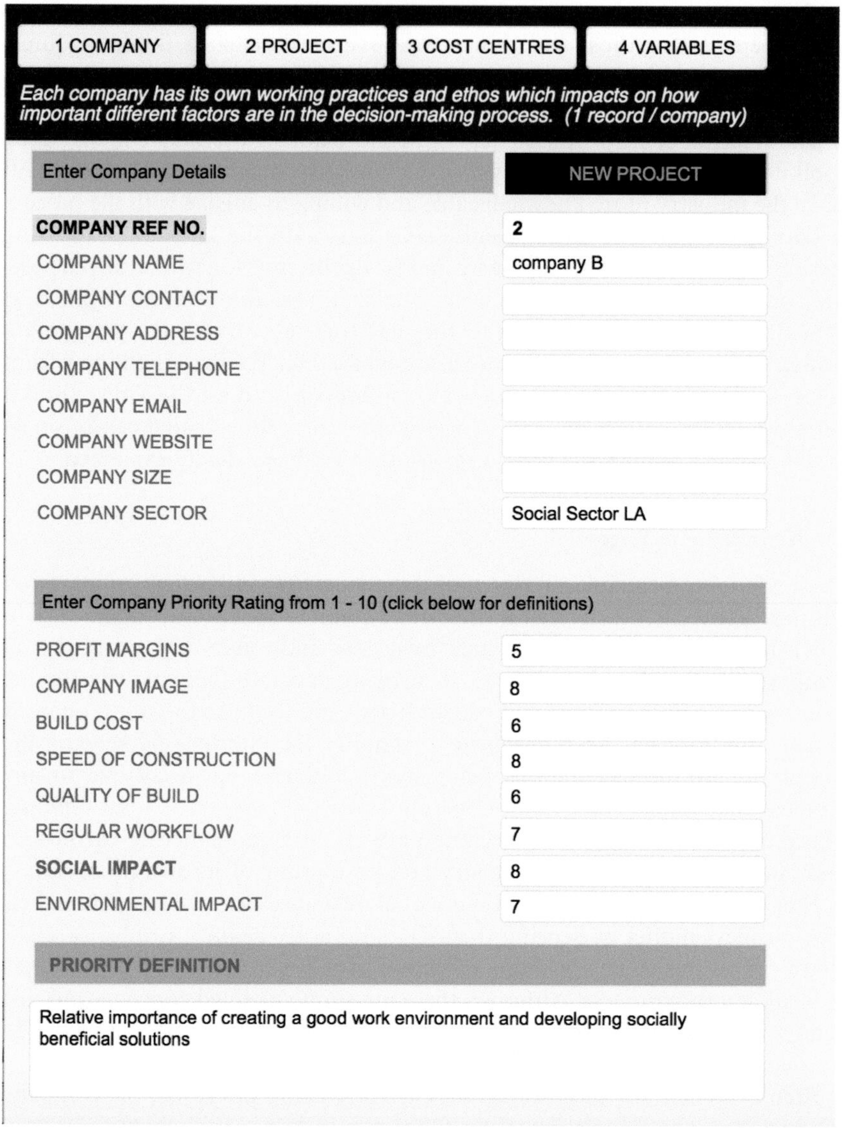

1 COMPANY	2 PROJECT	3 COST CENTRES	4 VARIABLES

Each company has its own working practices and ethos which impacts on how important different factors are in the decision-making process. (1 record / company)

Enter Company Details	NEW PROJECT
COMPANY REF NO.	2
COMPANY NAME	company B
COMPANY CONTACT	
COMPANY ADDRESS	
COMPANY TELEPHONE	
COMPANY EMAIL	
COMPANY WEBSITE	
COMPANY SIZE	
COMPANY SECTOR	Social Sector LA

Enter Company Priority Rating from 1 - 10 (click below for definitions)	
PROFIT MARGINS	5
COMPANY IMAGE	8
BUILD COST	6
SPEED OF CONSTRUCTION	8
QUALITY OF BUILD	6
REGULAR WORKFLOW	7
SOCIAL IMPACT	8
ENVIRONMENTAL IMPACT	7

PRIORITY DEFINITION

Relative importance of creating a good work environment and developing socially beneficial solutions

Figure 7.2 The Metriculator – Step 1, company profile.

with a series of pre-defined but adjustable industry assumptions that give a further weighting to different metrics. A metric that influences some aspect of the 'direct construction costs' at 60% of the total build cost, such as 'plant hire requirements', will clearly be of more relevance to an architect, a developer or a client than a metric that influences 'life-cycle costs' such as 'long-term

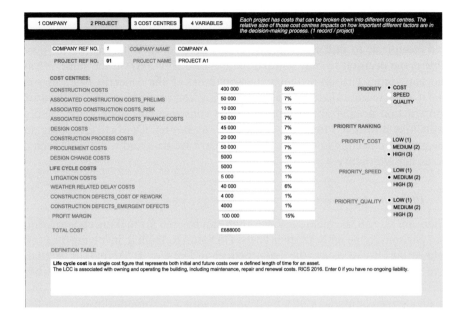

Figure 7.3 The Metriculator – Step 2, project costs.

maintenance' representing in this instance only 1% of that total. In contrast, an affordable housing provider with very different priorities regarding social value and long-term maintenance would find its metrics weighted to reflect their alternative business model.

How those algorithms have been constructed and the degree to which they can be manipulated will be explained later, but what is important is that the 'front end' that the industry initially engages with offers a straightforward qualitative assessment which answers many of the concerns found to be preventing businesses from engaging with the collating and sharing of metrics:

Profitability: *the cost impacts are linked to their own cost centres*
Relevance: *the impact of the metrics relates to their own business model*
Immediacy: *the data is relevant to the project in hand so of immediate use*
Security: *the data is collected for the company's own use*
Comparability: *the comparison is being made purely between different methods of construction with no other variables in play*
Simplicity: *the whole operation is based on existing in-house information*

What a business gets from this exercise is meaningful guidance on the possible alternative methods of construction available, based on the extent to which each of a raft of metrics represents a positive or a negative impact on a project's viability. Individual companies get information that is relevant to them and their perceptions of value, whilst government gets to see what those perceptions are – once anonymised – and where exactly they need to legislate and incentivise to shift them to where they need to be. What makes this possible is the decision to only compare alternative solutions on the same site at the same time. This immediately answers most of the concerns listed above, but it does require there to be a mechanism for comparing one method of construction against another prior to the construction taking place. Before explaining how that 'immediacy' challenge can be met, it's worth recapping on the options currently in use to gather comparative data and why this is such a necessary alternative.

The Alternative Approaches

There are three basic methodologies than can be used for making construction comparisons, none without their flaws: 1) Digital to Digital, involving desktop calculations to measure the hypothetical difference; 2) Actual to Actual, involving the gathering of data from two built projects on different sites, either current or historical; and 3) Actual to Digital, being put forwards here and involving the comparison of a built project with a hypothetical project on the same site.

After an industry-wide attempt to gather a body of digital evidence using the first of these methods, there has followed a wealth of case studies exploring the real-life data from built demonstrator sites using both traditional and modern methods of construction. The conclusions drawn from these studies however are invariably more about the problems encountered in collecting and interpreting the data than they have been about the evidence delivered.

To move this forwards, there needs to be a full assessment of the pros and cons of each of these first two methods, how realistic it is that the problems so far encountered can be overcome, and to what extent those problems are critical to the main purpose of the exercise being undertaken – that of changing industry mindsets. Whilst that represents a step back from where those promoting MMC or the Value Toolkit might want to be, it is fundamental to the success of those processes, as is the success of both MMC and the Value Toolkit dependent on each other's adoption. MMC solutions need to be seen as substantially more beneficial than traditional methods of construction before there will be any mainstream transition. Just being marginally preferable is never enough to unseat the status quo. Legislation and the mandatory use of procurement processes that favour more ecologically and societally beneficial MMC solutions are therefore important drivers to shift mindsets.

The other side of the same coin sees the implementors of the Value Toolkit needing a body of forward-thinking and innovative developers to engage with the process and embed their high-scoring solutions as the new standard to be measured against.

Method 1: Digital to Digital

Computer-generated models can accurately calculate some of the factors that need to be measured, such as an element's U-value, but others can only ever be estimated either due to the complexity of the design or the lack of complexity of the modelling algorithms. Other measurements, relating to those non-design-influenced factors, such as human, site and weather issues, become even harder to gauge, making the whole process hard to validate against real-life experiences.

Method 2: Actual to Actual

It is already clear that there are fundamental problems with the collection of data using the currently preferred method of comparing site A (MMC) to site B (Traditional). Any direct comparability is lost amongst a multitude of uncontrolled variables masking the true values being returned, on top of which there is the unrealistic expectation that a poorly performing traditional contractor will be able to gather, let alone have any incentive to gather, the information required.

Method 3: Actual to Digital?

If a digital to digital analysis only tells half the story, and comparing site A to site B cannot deliver meaningful comparisons, would there be any advantage in comparing the proposed method of construction on a specific site with a hypothetical alternative on the same site? Would this represent the best or the worst of both worlds?

We have already seen the advantages of this third approach, as it satisfies many of the conditions already discussed – simplicity, immediacy, security, and to some extent comparability. The limitations for this last condition are around the complications that only come to light during the actual build programme that cannot be easily allowed for in the hypothetical alternative. It is this weakness that is dealt with here by proposing an adaptation to this approach that takes the benefits from both the first methodology (Digital to Digital) and the second (Actual to Actual) to arrive at a solution that improves on the comparability of the third (Actual to Digital).

The key problems that arise from the 'actual to actual' site A / site B comparison method that are nullified by this third approach are those due to the differences in the location, site conditions, time of year, date of construction, variance in the buildings themselves and the companies involved – plus variations in the actual data-gathering processes used. Together these represent far too many variables to control in order to gather any meaningful data from comparable sites. In comparison, the only variables this third approach has to contend with, beyond the actual constructional differences that we want to compare, are the differences between the expected 'digital' outcome and the reality of actual 'built' outcome.

The term 'Digital Twin' is used to define the pre-production digital model used to test proposals and hone the design before building commences. That same Digital Twin can also be used on completion to compare the expected results with the built reality, providing valuable information, often referred to as the 'performance gap'. This is shown in Figure 7.4 as 'D1'.

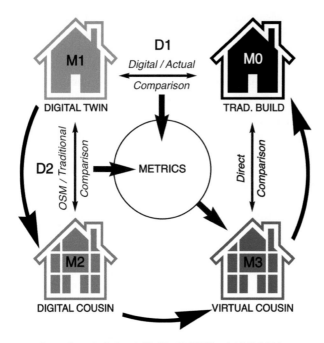

Compare the metrics for the actual Traditional Build (M0) against its Digital Twin
(M1) to measure the Digital / Actual difference (D1)
Compare the metrics for the Digital Twin (M1) against its OSM Digital Cousin (M2)
to measure the digital Traditional / OSM difference (D2)
Update the metrics for the Digital Cousin (M2) with the difference (D1) to arrive at
the metrics for the Virtual Cousin (M3)
Make a direct comparison between the Traditional Build (M0) and the Virtual
Cousin (M3) where M3 = M1 + D1 + D2

Figure 7.4 The Digital Cousin approach.

The proposal is to extend this principle and also define a 'Digital Cousin', a model of the proposed MMC alternative, identical in every sense to the Digital Twin, bar the constructional differences, which we've called 'D2', to be evaluated using the same metrics. The third step is to then combine these two differences, D1 and D2, to create a 'Virtual Cousin', a version of the Digital Cousin updated with the performance gap information, D1, gained from previous experience of the comparison between the Digital Twin and its final built form. That, as we will show, is a simpler process than it may at first appear, and results in this third model that can then be used to make a decision about how to proceed by comparing the projected costs for the MMC option M3 against the original proposal, M0.

So far this approach has addressed four of the six barriers to engagement – simplicity (proof coming), immediacy, security and comparability – by developing a system of comparison that can be internalised, using a business's own values and in-house metrics without having to divulge sensitive information to its competition, and doing this at a stage in the decision-making process where the findings can still be of use on the project being assessed. But to make those findings truly useful they have to be seen as relevant, which means there has to be some genuine financial benefit in collecting and acting upon this information beyond satisfying government requirements.

This is where we need to look at the inner workings of the Matriculator to understand how it personalises what these metrics mean to individual businesses. From the government's productivity-driven perspective, the metrics are necessary for two related purposes. Firstly, they are needed to prove the case for MMC in whatever form is beneficial to broaden our delivery base, increase the build rate, and improve sustainability to fulfil the net zero commitment. Secondly, and for similar reasons, they are needed to shift the industry's decision-making drivers away from immediate profit towards longer-term strategies based upon the Value Toolkit's broader definitions of value.

For the metrics to be seen as relevant to those collecting them, however, they must be providing information that is of value to those businesses directly, whether that be in informing a decision to transition to MMC or in adopting the Value Toolkit's approach. In either case, there must be some connection made between the long-term social and environmental values they are capturing and the short-term project cost centres they impact upon. This forms the basis for the inner workings of the Metriculator (Steps 3 and 4), where links are made between the 40+ metrics that are now in circulation and the most commonly used cost centres that will be recognised by most developers.

There are two points to reiterate before looking at this next, more complex level of decision making. These are inner workings, and not something that an end user would normally have to engage with. If they should want to satisfy themselves that the industry perceptions used to formulate the weightings were in line with their own sense of 'normality', they can do so, and any changes made would be recorded as part of the information-gathering exercise. The other point is that the way in which the weightings have been generated is based upon a simple exercise of direct 'more or less than' comparability. The simplicity of the comparisons is key to their universality. There are no conflicting methodologies that can interfere with a decision that is based upon whether or not 'programme certainty', for instance, is improved by the adoption of a different method of construction. Counter-intuitively, the less accurate the claim becomes, the higher the confidence levels can be in it being correct, and the more of these binary choices that are made to reach a final conclusion, the less likely it is that the conclusion will be wrong due to the omission of a critical factor.

Step 3 – So, for this next stage in the process, those 40+ metrics have been categorised into 15 that have a direct connection with known cost centres, such as prelims, design costs, etc. and those remaining, which have been linked back to these 15 core metrics where they can be seen to have an influence over them, effectively providing a weighting from those secondary metrics. A similar weighting process was then carried out on the 15 core metrics by relating them back to the company's business priorities from Step 1. This 'behind the scenes' complexity is what allows for the simplified user inter-action that is so critical for the process to be seen as 'adoptable'. There is nothing at that second level of functionality that cannot be adjusted, but like all intuitive interfaces, the division between first- and second-level interaction needs to be clearly defined.

Critically, a decision was taken at this stage to allow for more metrics to be added if seen to be missing. The metrics had first been categorised by where their impact could be measured, with relevance

- at a project level (such as programme certainty);
- at a company level (such as product accreditations);
- at an industry level (such as construction energy use); and
- at a societal level (such as community benefits).

The vast majority of the 'missing metrics' were found to relate to the impact of the changes being proposed at a company level, and most of these were risk related, highlighting the concerns that companies have about MMC that are not being recognised – such as flexibility of systems, disruption to working practices, maturity of supply chain, etc. Through adding these more practical measures and linking them all back to a company's cost centres and business model priorities, an increasingly accurate picture emerges of which metrics are seen as important and why.

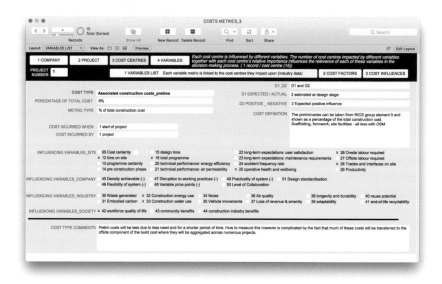

Figure 7.5 The Metriculator – Step 3, 'behind the scenes' metric influences.

Step 4 – The final step was to then hone the weighting given to all these metrics by adding a range of further options around three more factors:

- the levels of confidence in a metric (based on sample size, reliability of data collected, relevance of conclusions drawn);
- at what level a metric is seen to be relevant (project, company or industry); and
- when the information needed could be gathered (before commencing on-site, during, or on completion).

By way of example, a metric such as 'time on-site', which can be predicted and easily and reliably collected, is project related and has a direct relevance to production costs, would score highly compared to, for instance, a metric capturing air quality, which might be seen as hard to quantify, even harder to prove and only of benefit for future projects and at a societal level.

By feeding all these factors into an algorithm that weights the metrics before relating them back to the original cost centres that they influence, a hierarchy of importance can then be seen that is specific to the company using the process. This, together with the preferences entered about their company's own values, and the requirements of the job being considered, then allows them to make an informed decision about which method of construction will deliver, for them, the best results.

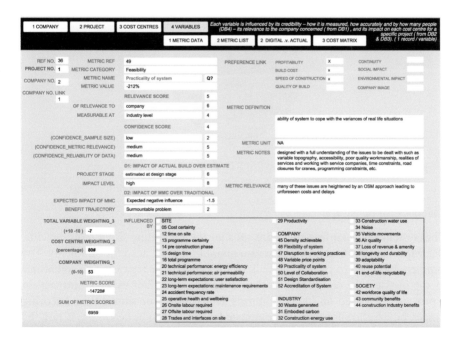

Figure 7.6 The Metriculator – Step 4, metric weighting.

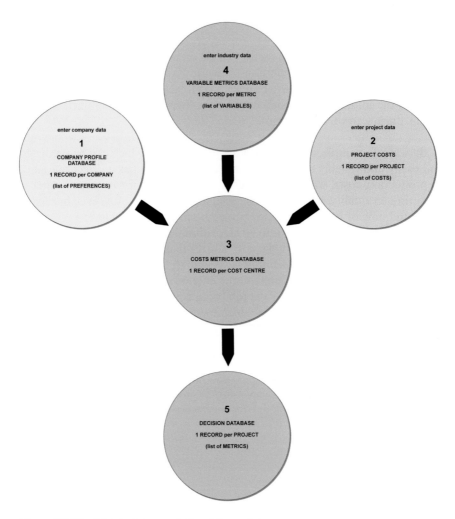

Figure 7.7 The Metriculator – relationship model.

The Appeal of the System

This system of weighting therefore satisfies the remaining three requirements, Relevance and Profitability, whilst also maintaining the need for Simplicity. The aim here has been to pre-set the bulk of the preferences regarding the relevance of the metrics, the confidence in their accuracy, and their expected impact on a project, etc. based on industry norms, using binary or simple 0 to 3 scoring ranges, but to allow for these to be adjusted if thought to be necessary. If not, the only data requirements from the user are the company values scores across eight categories and the estimated cost breakdown of the project across 15 categories.

In return, the user gets to see which metrics are important in terms of the influence they have over project costs, and which metrics have very little bearing, or even a negative impact over the decisions about to be made of which method of construction to choose. This is very different to a system that provides scores across a range of metrics for these different approaches which are supposedly accurate, but make no attempt to allow for the multiple variables in play and no attempt to determine the impact they might have on actual build costs.

The Argument Against

Playing devil's advocate, a necessary part of this process, the argument against this approach would be that it is attempting to monetise the metrics, against the spirit of the Value Toolkit's intention of redefining value to reflect societal and environmental benefits alongside the usually dominant cost benefits used in the decision-making process. But the reality of how the construction industry currently operates and how difficult it has been to 'shift the dial' requires a more pragmatic approach to be adopted. 'We have to start with what we've got to get to where we need to be' – which is another way of saying that we have to understand the barriers to adoption from the perspectives of those we are expecting to engage with this process before we can know how to influence their opinions. A major requirement of this approach, however, in order to shift that dial, is in explaining where the hidden benefits are in adopting a new approach that may seem to only deliver collective benefits. This might be as prosaic as better access to funding, finance or government contracts, or in reduced maintenance costs, or futureproofing against retrofitting costs – but all of these are ways in which longer-term societal gains can be shown to have a shorter-term financial value, and become, in some cases, immediately relevant at that individual business level.

And finally, in addition to having that greater impact, what this approach would deliver through allowing users to engage on their own terms is the evidence the government needs about how the industry is currently thinking at a granular level. With finite resources, this is what is needed for government to focus its attention where it is most needed, and also how a proposal such as this would need to be portrayed to be appealing from their perspective.

The Metriculator's Chances

As a questions-based tick-box approach, that's all six questions answered satisfactorily, which in theory should result in an easy path to adoption with no major barriers left to deal with. But someone or some industry body still needs to promote the solution up to the top of the tree, before then promoting it back down to the entire industry including its 400,000 hard to reach SMEs. Plus, the industry has to have the bandwidth to take this, along with the many other initiatives being rolled out simultaneously, on board. It seems ironic that, whilst the one overriding request being made of the construction industry is to pull its fragmented parts together and act more collaboratively, these initiatives, funding programmes and legislative upgrades that are constantly being dropped into its mix couldn't be more disjointed if they tried. With that in mind, the next proposal is an attempt to simplify the bigger picture by pulling some of those initiatives into a more accessible structure.

The Supply Chain Database – a Solution to Siloed Thinking

Fragmentation exists throughout the whole industry, not just amongst the supply chain companies. For all companies, accessing advice and information about policy, funding, legislation, accreditations, etc. can be challenging, with

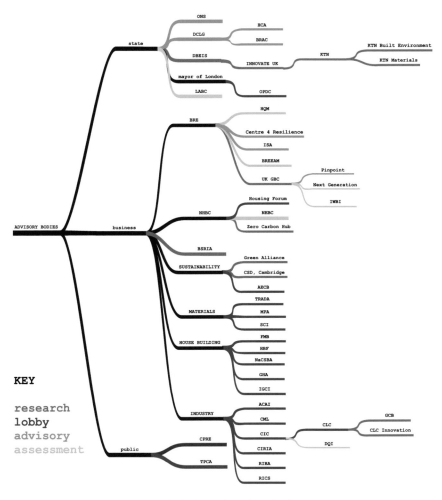

Figure 7.8 An attempt to map the industry's various bodies.

multiple bodies operating in similar spaces with ill-defined overlaps resulting in considerable duplication of effort, both for those providing the information and those seeking it out.

For small SME companies with little surplus resource to allocate to research, data gathering or 'non-essential' decision making, prioritising anything that isn't mandated is unlikely, and going further than is needed to ensure compliance, equally so. It is therefore understandable how so many well-meaning initiatives fail to get the traction they feel they deserve, and even when they do, find themselves being complied with at a superficial level that avoids businesses having to engage with the true spirit of the changes being imposed.

This proposal goes beyond ensuring that these initiatives are promoted in a way that makes the benefits to those companies clear, and hits upon the even more difficult task of making those thousands of small SME companies prioritise even considering, let alone adopting, new ways of working over the more pressing day-to-day demands on their time.

What all of the current slew of government initiatives have in common is the fact that they are all primarily, if unintentionally, aimed at the 'top end' of the supply chain, the larger delivery companies with a close or direct relationship with the government procurement processes where compliance is easier to enforce. But what they also have in common is the need for the rest of the supply chain to play their part for them to be effective. But the further down the supply chain one travels, the less control there is over that decision-making process, and the greater the need for coercion to be tempered with persuasion through a better explanation of the benefits to be gained.

Sat within the three-stage process defined earlier when looking at how to guarantee support for the Platform Design Approach (Figure 7.9) are three of those initiatives: LEXiCON, the Route to Compliance, and the Value Toolkit, each having a vital role to play, each developed through the Building Research Establishment (BRE), but all being dropped on the industry as independent initiatives. Without an overarching structure in place to help the industry see the bigger picture for how these tools all work together, and the ability to access them through a single interface, it is unlikely that all, if any, of them will receive the attention they deserve.

LEXiCON: Machine-readable 'Product Data Templates' aimed at standardising construction product information to support manufacturers in sharing product information freely across the industry.

Route to Compliance: A comprehensive database of standards, legislation, and testing methods for all construction industry processes and procedures (currently shelved).

The Value Toolkit: A tool designed to help built environment organisations better define, understand and manage the value created throughout the life cycle of a project.

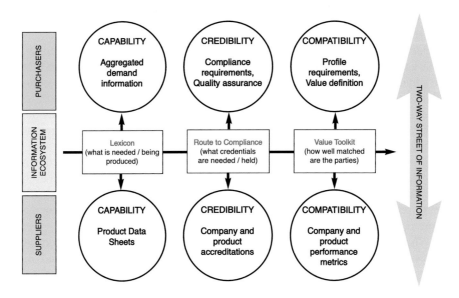

Figure 7.9 The three steps of Capability, Credibility and Compatibility.

Even at that 'top end' where compliance is mandated, clarity of purpose and the benefits to be accrued from adoption are important ingredients to ensure more than a superficial level of engagement, but the further down the chain and the smaller the business, the more important this becomes.

How this interface with the industry is structured therefore becomes critical to the success of all these programmes, with a similar list of criteria as before that need to be satisfied: It needs to give the industry confidence that this is not just another passing panacea to pay lip service to before being allowed to return to business as usual, and it also needs to reflect the relevance of the initiatives to each of the many different sectors within the supply chain. To achieve the former, there needs to be ownership at an industry level to prove that there is genuine ongoing support and also ensure there is only one definitive source for the industry to deal with. To show relevance, however, the message must reflect the many different perspectives that are being considered, and therefore needs to be translated for each of those stakeholders, with access for each through the appropriate portals.

There are many industry bodies that will need to perform their part in this role, with their websites providing the entry point and the route through to a central resource, but with the appropriate wording, case studies and explanations of how to approach the relevant parts of the overall process. This also allows for the 'scale of engagement' to be managed, with certain sectors only required to get involved at an entry level, or with only a smaller

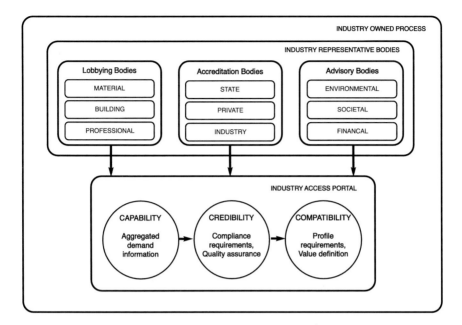

Figure 7.10 One central resource with multiple entry points.

part of the entire process. Proportionality is an essential factor to ensure processes such as these work for all projects, irrespective of size or complexity, so that the time and effort required to engage is always seen as being appropriate.

By way of example, a small SME timber frame company might initially find out about these initiatives through their membership of the Structural Timber Association (STA), whose website would be one of many portals through to the central resource. The STA would want to inform their members of this and would have set up a two-way link between their membership database and the national LEXiCON database as a mutually beneficial exercise. The STA site would explain how timber frame companies could benefit from this by agreeing for their company information to be linked through to a central resource that could be accessed by potential clients. Control of that data and whether it can all be seen immediately or be partially blocked until a direct request – or even a payment – is made would remain with the timber frame company.

They might then receive a request for that information from a housebuilder wanting them to quote for a roof truss contract. Before agreeing to enter negotiations, the timber frame company could then find that housebuilder on the same central resource to check their status, their potential for ongoing work, any accreditations held, etc. That might reveal them to have BOPAS accreditation, for instance, information held in the system because of a similar two-way link with BOPAS's database of their clients.

After establishing that the product is appropriate and that the timber frame company has the credibility and capacity to supply the trusses in the timeframe necessary, the link through to the Route to Compliance database could then be used to check whether any specific accreditations or assurances required to tender for the contract are held or in place. This might reveal that the timber frame company's insurance cover is inadequate, or that it isn't signed up to a particular framework, but rather than these being barriers to starting a conversation, they become conditions to be met should the contract proceed. Again, two-way links with these bodies provide both that centralised source of information and also a link back out to each body to arrange any further cover or agreements needed.

Before spending time and resources on meeting compliance requirements, however, the third step can also be dealt with, where the two companies assess their compatibility under the Value Toolkit's requirements, which might be an internal agreement between the two companies or one dictated by the housebuilder's client, for instance. So far, none of these are onerous demands on a small SME company if contained within one defined process and dealt with in the right order and with the right levels of control over the data being shared. Utilising a company's existing data at the outset of negotiations means many more companies can enter the ring at a far lower risk to their immediate profitability. Requiring potential clients to pay even a nominal amount for any further information needed that might require time and resources to gather or measure reduces the barriers to engagement even more.

The ongoing financing of such a resource should in fact be easy to achieve. Between the many entry portal bodies, all of whom receive membership fees, the upkeep of a system that operates off mutually beneficial database and website links would be a negligible shared cost. Access to the resource would be entirely through these bodies, making it both controlled but available to any business that belongs to at least one of them.

Currently, none of this is being considered because it sits outside the scope of the teams responsible for developing each of the component parts, even though those teams were at some point all connected through the BRE. Teams come and go, and projects get shelved for a lack of continuity or connectivity, with the Route to Compliance database being one of those casualties. Being allowed to step outside the scope of any project to look at the environment it sits within and test its veracity from that broader perspective is what a Systems Thinking approach offers. Ultimately, it requires the entire industry to think more collectively, and not be constrained by the need for immediate results to the detriment of longer-term impact. This final overarching tool is therefore an attempt to at least steer the industry in that direction.

The Decision-Making Tool – a Solution to Compliance-Led Thinking

The Supply Chain Database was an attempt to overcome siloed thinking by tackling all three of the barriers preventing change in the construction industry: knowledge, motivation and ability. How does an entire industry get to know about an innovation in the first place, why would they feel it was worth their while engaging with it, and even if they did decide it was worth their while, how feasible would it be for them to adopt it? These are the three basic questions that need to be asked of any innovation to assess its chances of success, and they apply equally to both the innovator and the prospective adopters of that innovation. And, as stated earlier, this also applies to a process or a concept such as a decision-making tool in the same way as it does to a physical product. There is a certain prophetic quality to the argument that says 'if such a tool fails to get itself adopted, it would have probably failed in its purpose of getting innovations adopted too'.

The barriers to getting a whole industry to think differently have been covered now across a range of proposals that have all had to address similar problems. This is now about how to encapsulate all of that into a question-based decision-making tool that can and would be used by enough people to make a real difference. To be clear, we're talking about a digital, database-driven tool that would need to be internet based but accessible through any device.

At its most basic, this approach is not about disseminating the right answers, but about helping the industry understand what the right questions are that need to be asked:

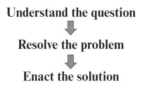

Understand the question

Resolve the problem

Enact the solution

The question initially being asked is rarely the one that needs answering. Fully understanding what the question is and why it is being asked is therefore the first step, before then gathering all the known information so that the gaps in the knowledge needed to resolve the problem can be identified. Only once those gaps have been filled and the knowledge held is complete can the options available be drawn up and an informed decision be taken on which way to proceed.

The route to identifying the root cause that lies behind a stated problem – the issue that really needs resolving – is through asking the right questions, and then continuing to question the answers given until there are no areas of doubt left to question. This is true for all problem solving, and establishing the validity of a new innovative solution is no different. For this to work as a process that can help to establish the worth of an innovation, and do this quickly and efficiently, there needs to be a structure to the questions being asked.

The starting point is to just brainstorm as many relevant questions as possible, irrespective of whether or not they can be answered. Then they can be structured into categories, and then they can be interrogated from all different perspectives, looking first at questions around the benefits that could arise, and then questions about the barriers that would need to be overcome. Those perspectives, many of which are so often ignored, fall broadly into five categories:

- the promoter of the innovation;
- the proposed adopter of the innovation;
- other industry sectors with an interest in the outcome;
- government sectors with policies to enact; and
- society as a whole.

This is an exhaustive process – as opposed to exhausting – but it quickly exposes the barriers to be removed or avoided and also where support can be found and collaborations made. It is basically an extended SWOT analysis that 'pigeonholes' the information gathered as discrete facts broken down to a level where they can be translated into binary choices that are far easier to make before settling on a strategy to enact. The entire process is captured in Figure 7.11, but is easier to appreciate through some examples of it in use. The two chosen here are the Retrofit Conundrum and, to prove that it can even be used to promote itself, the Decision-Making Process.

Example Question 1: Housing Retrofit

Why is housing retrofit not happening at the scale or pace required?

The Known Issues:

Not happening quickly enough
Too expensive
Not enough skilled labour
Every house is different
No regulatory guidance
Inconsistent funding programmes
Mixed tenure housing
Payback periods too long
No interest from mainstream builders
Not driven by the public
No clear technical solutions

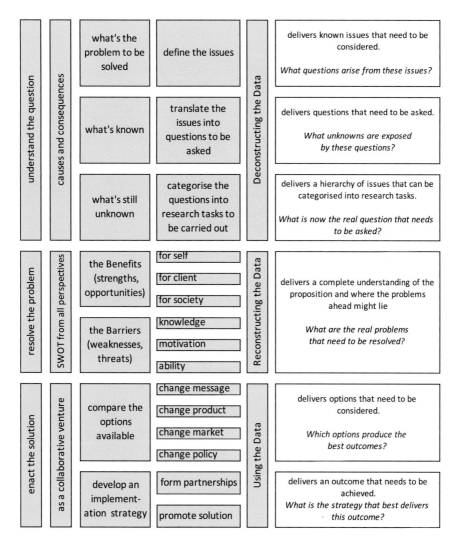

Figure 7.11 Stages in the decision-making process.

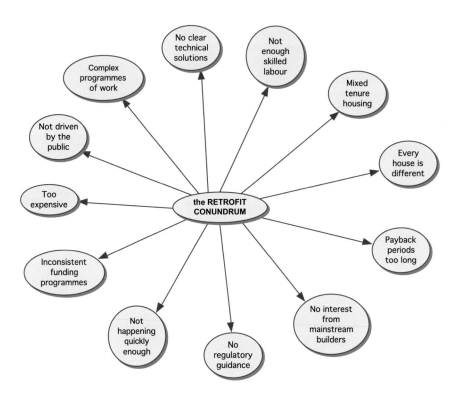

Figure 7.12 Stage 1 – identify the known issues.

Step 1: Understand the Question

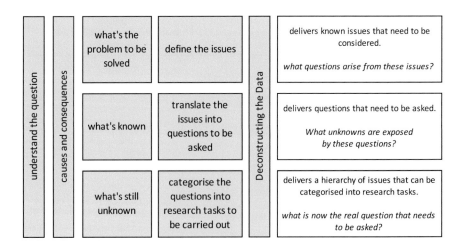

To visualise this more clearly, these issues first need to be organised into causes and consequences, which in this instance puts everything except the question being asked above the line:

But there is always a hierarchy to these causes, which then categorises them in a way that makes the challenges easier to define:

As patterns begin to emerge, it becomes clear what the main issues are and why it is so important to keep questioning the causes, especially those presenting themselves as root causes at the top of the map. By doing this, more causes and consequences tend to emerge, opening up new lines of enquiry to be followed up until all the 'whys' have been answered.

Finally, by sorting the evolving mindmap into a clear hierarchical order in Step 5, it becomes possible to draw some early conclusions:

- The current state of play is due to a combination of 'Retrofit Realities' (it's complex, expensive work with only long-term benefits) and 'Political Choices' (how to tax retrofit and energy, the repercussions of Brexit, Right to Buy, etc.).

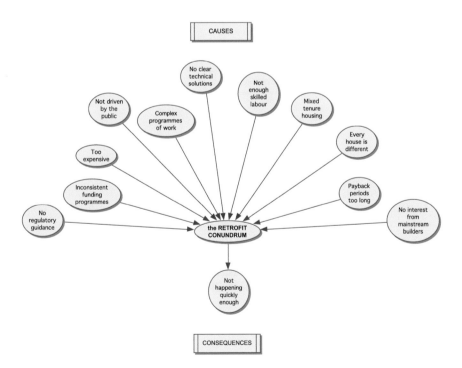

Figure 7.13 Stage 2 – causes and consequences.

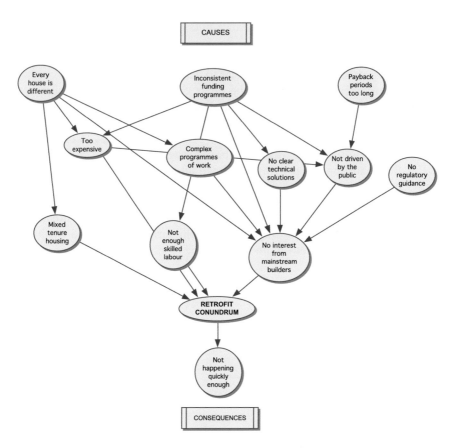

Figure 7.14 Stage 3 – hierarchy of causes.

- The combined impact of these is a mainstream public who are not driving retrofit, and a government not investing in the skills or infrastructure needed to create the right environment for it to thrive.
- As a consequence of this there is little incentive for the mainstream construction industry to move into this sector, making any work that is carried out even slower and more expensive than it should be.
- The slow progress resulting from this means we will not meet our net zero targets, more homes will slip into fuel poverty and pressure to demolish existing underperforming properties will build.

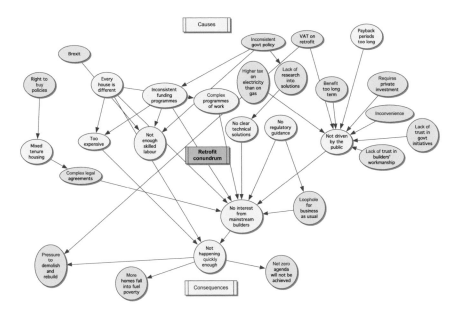

Figure 7.15 Stage 4 – the addition of further causes.

Step 2: Resolve the Problem

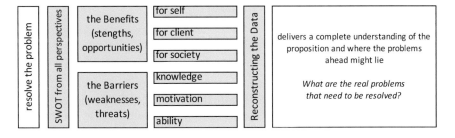

The most connected issues highlight where the most effective intervention points would be in the hierarchy of causes and consequences, with the option then of whether to accept, remove or avoid the barriers to progress they represent. The Retrofit Realities, as their title suggests, are barriers to accept, whereas the Political Choices could be influenced through a long-term, collaborative campaign. The two main targets to focus on, however, would be the public's and the mainstream industry's lack of engagement, and the multiple reasons that sit behind both of these.

This now becomes a systematic process that looks first at the benefits to be gained from retrofitting, which for the general public would involve looking in

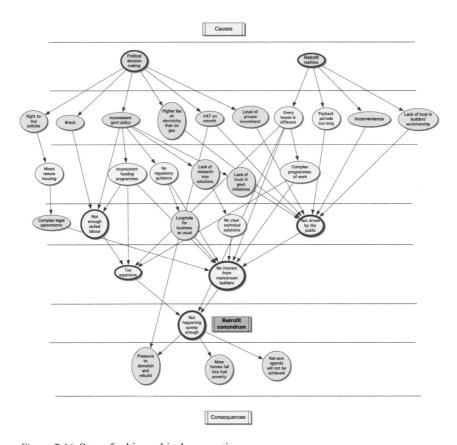

Figure 7.16 Stage 5 – hierarchical connections.

turn at individual occupiers, the landlords of rental housing or social housing, and then the broader societal gains, before then looking at the barriers to retrofitting in terms of knowledge (what to retrofit), motivation (why to retrofit) and ability (how to retrofit). Similarly, repeating that exercise for the mainstream housing industry, what would be the benefit to them in moving into this sector – as individual companies, as the long-term owners of the properties, or for the industry as a whole? Asking these questions of the industry, however, means that, when looking at the barriers to engaging, this must be done from the perspectives of all the other sectors involved in the process other than the end users – the builders, the project managers, the suppliers, the warranty companies, the design companies, the financiers and the assessors. It has to be a systematic, but exhaustive, process.

And what this process unearths are many differences of opinion. It exposes conflicting motivations that might create barriers to progress, but it also points the way to collaborations where similar outcomes are sought by different stakeholders. The benefits to be gained by the general public from retrofitting, for instance, could be said to be any of the following, and cannot be assumed or generalised:

- increased comfort levels;
- reduced energy bills;
- increased property value;
- lower personal carbon footprint.

The barriers preventing them from proceeding, on the other hand, could be any one or all of these:

- lack of knowledge of the potential benefits;
- frustration with complex funding routes;
- lack of awareness of our climate emergency;
- payback on investment too long;
- cost of works not reflected in increase in property price;
- incentive schemes not tempting enough;
- legislation not driving industry to deliver;
- lack of faith in quality of workmanship;
- lack of peer pressure;
- not intending to stay in property long enough;
- no access to private funds;
- no access to loan facility;
- not willing to lose storage space for intervention;
- inability to deal with disruption during works.

That is a far longer list of 'cons' than 'pros', and begins to explain why there is little demand coming from the general public for retrofitting, even when it is heavily funded. A developer's list of pros and cons would be very different, however, and would represent a far simpler transactional assessment of the situation around 'what funding is available?' and 'what are the criteria that must be met for compliance?' None of the public's motivators would be of much interest to typical developers as they have little interest in a property's performance after the contract is complete, and know that the public might claim an interest in their home's performance, but will rarely be prepared to pay enough to cover the cost of these unproven and less tangible benefits. So if government funding is not going to drive this, the viable retrofit options become fewer. A local authority or housing association on the other hand would be able to recognise and financialise the longer-term benefits of less homes in fuel poverty; increased housing stock value; longer property life expectancy; job creation, etc. – whilst fulfilling a manifesto commitment, making this a more plausible market.

Step 3: Enact the Solution

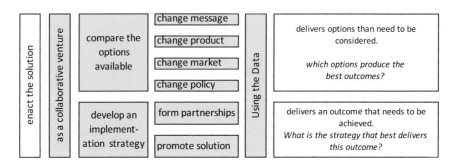

This more nuanced analysis of the market leads on to the next step in the process, which is to now analyse the options available for removing those barriers and how appropriate they may be in these different instances. Those options can be broken down into four categories of action:

- change the message;
- change the product;
- change the market;
- change the policy.

Which route is chosen depends on all the factors being considered, including which sector is being asked the question.

For the public end user or occupier, the options would be:

- *change the message:* work on behavioural change and education around the broader benefits of retrofitting rather than just focusing on 'what to do'
- *change the product:* find less disruptive / more affordable solutions that are less contentious and easier to promote – such as whole house infrared heating?
- *change the market:* focus on social housing and households in fuel poverty only, rather than trying to retrofit the UK's entire housing stock
- *change the policy:* improve finance terms, change tax regime to encourage retrofitting over new build and electric over gas.

For the developer, the options would be:

- *change the message:* focus on the counter-cyclical benefits of retrofitting as a source of work that could smooth out the peaks and troughs of economic swings
- *change the product:* develop more standardised packaged solutions that can be easily rolled out at scale across regular housing types

- *change the market:* don't focus on OSM providers where factory overheads require them to continue building during recessionary periods, but instead focus on traditional builders who get laid off
- *change the policy:* promote better focused funding schemes, changes to the tax regime and investment in skills training for traditional builders nearing retirement.

These are not either / or choices, as some represent long-term strategies that need to be set in motion whilst other quick-response strategies can begin to have an immediate impact. The important condition is that all strategies must be aligned, with the short-term actions working towards the longer-term goals, and vice versa, for the long-term strategies to define those goals and incentivise the industry to take small steps in that direction. All of the above approaches are in fact complementary and necessary, making for a raft of policies to be worked on and implemented concurrently.

That implementation strategy is the conclusion of Step 3 in the process, where synergies are used to define where true collaborations can be found and utilised. This is of particular importance for government, who cannot be seen to be favouring any one sector or industry over any other without having to compensate for this in some way at a later date. So again, solutions that benefit all sectors are therefore preferable to ones that only work from one perspective. The following four examples show how implementation needn't be a divisive process that pits one solution against another, but can instead help generate the collaborative environment needed before any innovation can flourish.

Traditional Housebuilder vs. Off-Site Manufacturer Business Models

New build and retrofitting would be better looked at in unison. To make full use of our limited skilled workforce it does not make sense to drive a retrofit campaign at the height of a housing boom, and neither is it an efficient use of skilled labour to lay the bulk of that workforce off during a downturn, as suits the traditional housing market's business model. The OSM industry on the other hand has invested in plant and machinery and needs to maintain a steady throughput of work across all economic cycles to be cost effective. The OSM industry is also better suited to the needs of the social housing market for reasons already stated, which means the social housing sector is best placed to finance that continuous demand through recessionary periods, whilst the skilled labour laid off by the volume housebuilders should be utilised to build a counter-cyclical programme of retrofitting work, where their broad traditional skills will be more appropriate than those of a newly trained OSM workforce. In addition, those retiring from the traditional build sector should be encouraged to lead those retrofitting training programmes and pass on their traditional skills where they will be most needed.

Local Authority vs. Central Government Decarbonisation Roles

Over half the UK's local authorities have declared a climate emergency and announced net zero agendas that require them to fully decarbonise up to 20 years earlier than the CCC's proposed ambitious but binding target of 2050 (Environmental Law Foundation, 2021). That date was arrived at by calculating the time and investment needed to prepare the way for a net zero economy, through building infrastructure, training a skilled workforce and bringing forward the necessary legislation. The reality of going ahead of that national legislation is that it can generate conflict more than collaboration, with private developers understandably resorting to legal defence to protect their profit margins. To avoid that conflict, whilst taking advantage of a well-meaning groundswell of public opinion in favour of decarbonisation, local authorities need to be funded to prepare the way for the national legislation to come through education and behavioural change programmes to reduce energy use rather than energy demand. Retrofitting properties and generating energy at an individual occupant level to reduce demand on the grid is costly in terms of financial outlay and less efficient in terms of embodied carbon. It will be necessary, but to be effective it must be done as part of a national strategy, not a regional beauty pageant to see who can pretend to get there first.

Heat Pumps vs. the Alternatives

One of the few things the UK can truly boast about as being 'world beating' is its gas distribution network. Had we embarked upon a hydrogen programme 30 years ago, so says the hydrogen lobby, we could have developed a blue, leading to a green, hydrogen industry and converted all UK households from North Sea gas to hydrogen without the need for air source heat pumps and all that comes with a complex and costly transition to electric heating. In reality, all transitions will be complex and costly, we will need many solutions, and government must decide which to use where. Heat pump technology can be made to work in new build properties now and needs to be honed for that market, in some situations by considering exhaust air heat pumps as a more practical alternative. But heat pump technology is rarely a viable solution for retrofit, so we need a realistic alternative that does not require the same levels of insulation, airtightness and the consequent need for bulky mechanical ventilation systems. That could be hydrogen in the future, if the distribution network could in itself be retrofitted to 'contain' the gas's smaller molecules, or it could be something completely different such as far infrared fabric. A quick-response heating system that primarily heats surfaces and the occupants rather than the air itself has multiple benefits in terms of health, less stringent airtightness targets and therefore less need for mechanical ventilation. There will be new technological solutions and they will upend our thinking, which we must remain open to. The immediacy of infrared heating will, from a behavioural perspective, beat the slow-response heat pump technologies every time, as will the simplicity of its application over the complex retrofitting operations currently being trialled alongside relocation programmes. Simple installation processes and quick-response heating solutions are just two of the factors that are not being given enough weight due to the failure to consider the options available from all perspectives.

Net Zero vs. Fuel Poverty as Retrofit Drivers

Energy security, affordability and carbon neutrality may be aligned as long-term goals, but in the immediate future they are often in conflict, as are the government departments responsible for their delivery. The new Department for Business and Trade, Department for Science, Innovation and Technology, and Department for Energy Security and Net Zero, together with the Department for Levelling Up Housing and Communities and the Treasury, all have their agendas and short-term goals to consider. With finite resources come difficult choices, but their collective choice has to be how to reduce carbon emissions affordably without endangering our energy security. Is retrofitting every property in the UK necessarily the best use of that finite resource, especially when decarbonising our entire energy supply through the development of our own renewable sources could be far less expensive?

These are all issues to be debated in their own right, but they are also relevant to how we solve the problem of retrofitting our housing stock and the role retrofitting should play in the ultimate goal of managing climate change. Nothing is unrelated, and any decision that fails to recognise the broader consequences of the action being taken, unintentional or otherwise, will be of limited long-term benefit. These are some of those consequences, from national to personal, that we ought to be considering before launching a national retrofit campaign we cannot afford:

- The payback period of the embodied carbon associated with the measures being proposed. If this is longer than 25 years, it will have no impact on a 2050 net zero target.
- How should that payback period be calculated? Should it be based on the carbon footprint of our national grid now or be based on an average of its projected footprint between now and 2050?
- Are we diverting limited resources to local climate emergency driven campaigns that would be better spent on a national effort focusing on longer-term solutions?
- If we can manage to fully decarbonise the grid by 2050, is there any point in retrofitting all homes to such an extent?
- How will the savings accrued by occupants due to lower energy bills be spent? If this is on an overseas holiday, for example, the carbon benefit will be immediately lost.

The point being made here is that the only way to align the campaigns around net zero, fuel poverty and the affordability of our housing is to accept we cannot realistically retrofit all of our homes, and start with those most in need in the hope that before too long we'll have developed a solution we can afford to roll out beyond this.

Example Question 2: The Decision-Making Tool

How can an entire industry be made to use it?
The same approach must now show how this decision-making tool can be used to successfully promote its own use. If it can't, it's unlikely it will ever get the chance to prove itself in any other way. The starting point is the same, in that we need to understand the question being asked and then deal with the three barriers to adoption – knowledge, motivation and ability as before. But first, the need for a decision-making tool needs to be questioned. Do we really need one, and if so, why?
The industry's poor track record in terms of adopting innovative solutions or working collaboratively to increase productivity levels is evidence enough of the need to improve its decision-making process and save itself from an existential collapse, but we also need to understand why the industry has arrived at this impasse before we can improve upon it.

The Known Issues:

The possible causes and consequences of the construction industry's poor decision-making process are many and varied:

- Industry always pushed for time, always short of resources.
- Industry very fragmented and adversarial, working in silos with little collaboration.
- Decisions often taken with only a fraction of the necessary information to hand, as a consequence of the above.
- Decisions often skewed by one sector's influence / control over the process, such as shareholders' dividends, government regulation, warranty industry's conservatism, etc.
- Decisions increasingly led by compliance requirements rather than a full understanding of the need or purpose.
- Embedded knowledge at a personal or business level acts as a barrier to new ideas being adopted.
- Cyclical nature of the industry means little emphasis put on long-term strategies, R&D, training, etc.

- Pace of change outstrips the industry's ability to fully trial and test new solutions before implementing them at scale.
- The time taken to make decisions tends to be fixed, regardless of the complexity of the problem, resulting in poor choices being made.
- Short-term needs will always trump longer-term aspirations – bottom-line profit over societal and environmental benefit.
- Slow uptake of innovative solutions due to lack of knowledge of their existence.
- Poor understanding of risk and a lack of resources to measure its true impact.
- Reluctance to share knowledge and experience with competition.
- Tendency to stick with what's known to work for an easy life.

As before, these issues can be structured to show a pattern of causes and consequences:

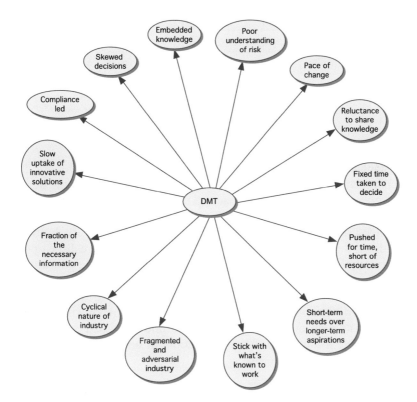

Figure 7.17 The decision-making tool – known issues.

And once added to, after further enquiry into these issues, the similar top-level root causes emerge of political, economic, societal and environmental factors over which we have little, if any, immediate control. The factors that emerge as most connected however are the ones to focus on. The pace of change and daily pressures in the workplace often result in a culture of quick decision making with little time to fully research the problems being confronted. And a fragmented and adversarial industry prevents the collaborative sharing of knowledge, which in turn slows down the adoption of BIM, exacerbating further poor, siloed decision making and a consequential reluctance to risk transitioning to modern methods of construction.

So if that's the problem, how best can it be resolved? Who stands to benefit and who would want to, or inadvertently resist an improvement to how we make decisions?

Resolving the Problem

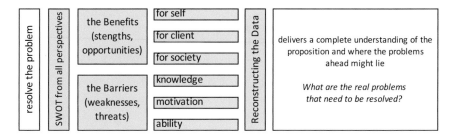

We've established at least some of the reasons why we've got such poor decision making, and through that some of the areas where changing the way the industry operates could have an impact on this: The widespread adoption of BIM would help to facilitate information sharing between different sectors, but the siloed state of the industry does not make that an easy fix. Neither is slowing down the speed at which decisions have to be made a realistic option. These structural realities appear to be just as immovable at the political and economic forces that have created them, making it difficult to see any obvious entry point for where an innovation such as this could be seeded.

As with BIM, the collective benefit of better decision making is clear, both at a business and a societal level – the immediate and individual benefit less so. But this is where it must begin. Any innovation, or any change to the status quo, be that to product or process,

- must offer a substantial improvement over whatever it aims to replace, and be able to do this for all eventualities;
- must benefit all those involved in the process from beginning to end;

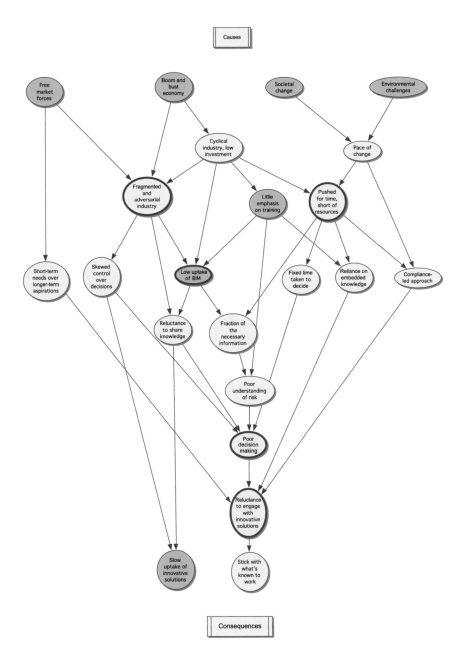

Figure 7.18 The decision-making tool – causes and consequences.

- must do this for less money, or in less time or with less effort, and preferably all three.

For the individual charged with having to research a problem, the critical factor would seem to be the speed at which a solution can be arrived at, even more than how effective the process is at arriving at the best outcome – at least at the outset. Over time, poor decisions take their toll, and the benefits derived from making better choices become more apparent, but if the time or effort involved in understanding and applying the process is too great, that next level of realisation will never be reached.

The process therefore needs to be simple to understand and quick to apply, regardless of the scale of the task it is being applied to. Too often solutions are designed to work for a specific set of conditions and are not viable outside of those parameters. This is as relevant to a decision-making process as it is, for instance, to an innovative foundation system. Until a solution is developed that has the flexibility to work at less cost and cope with all eventualities, it will not be a contender. For a decision-making process such as this, cost is not necessarily an issue, but speed and simplicity are, not just in reaching a conclusion, but in justifying how that conclusion was reached.

A systematic process where the steps are laid out as a series of concatenated questions to be asked, but with the option to only drill down and answer those that are relevant, begins to suggest how a reliable, exhaustive process could be developed that not only captures and 'pigeonholes' any information gathered, but automatically produces a structured report that follows a logical sequence of decision-tree steps towards a final conclusion.

Now we have a reason for both management and researchers to comply with a system that allows for a report to be methodically developed within an agreed framework that delivers a traceable decision-making process back down from a one-paragraph report for the CEO, through an executive summary for publication, to the original research for referencing. But if those are the benefits covered for businesses, clients and the wider society, what are the barriers to such a system's adoption?

Knowledge, motivation and ability. Motivation has been covered, as the benefits have been established, and unlike BIM, this is not a change that is only of benefit if everyone agrees to adopt it. Any one business can benefit individually, as long as there is 'buy-in' across all within that business, so in that respect this should be a much easier innovation to promote. The ability to adopt something of nominal cost is also unlikely to be an issue, but knowledge of such an innovation, and proof of its capabilities, is still a major barrier to overcome. The main problem is therefore who is going to promote a decision-making tool that could effectively be free, or only a nominal cost? How can the solution, once proven to be beneficial, be publicised and promoted?

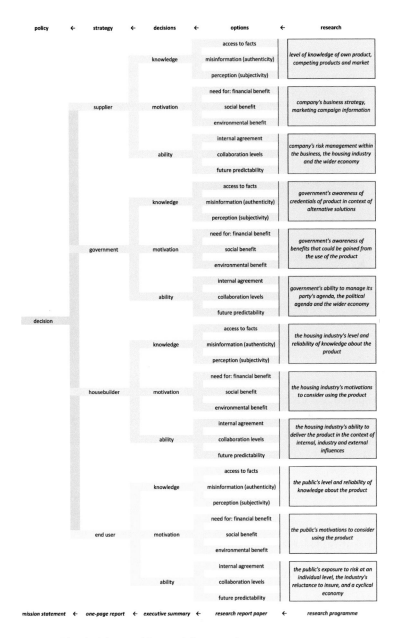

Figure 7.19 The decision-making tool framework.

Enacting the Solution

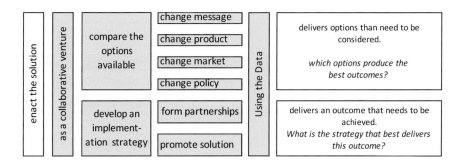

As before, if the barriers to adoption are still too great, the options are to either change the way it is being promoted or the market it is being promoted to, or change the product until it is more appealing to the market, or change the policies that might be unfavourable to it being seen as necessary. And again, these are not either / or options as it is often necessary to consider all four approaches either in unison or as part of a longer-term strategy.

Change the Product

If developed as an opensource product, the data and knowledge gathered could become a shared resource and function under the 'Wikipedia for Construction' model, ensuring that the data is updated and maintained, with access paid for through a licence fee.

Change the Market

All businesses across all sectors within the broader construction industry have to make complex decisions and would theoretically benefit from such a process. Creating points of access for different sectors would however allow for different perspectives to be captured and viewed separately and act as an educational tool for each sector to better understand the motivations of others. It could also allow for a sequenced roll-out, focused on one sector at a time.

Change the Message

This is a decision-making process but that is not necessarily how it needs to be promoted. The outcome of better decision making is higher productivity and profitability, which is what business leaders will respond to. For those carrying out the research, on the other hand, freedom to arrive at and justify their own conclusions through a structured process that saves time is a far more powerful message. As an innovation that can be equally introduced from above or at a grass-roots level, and preferably both, getting the right message for the right audience is essential.

Change the Policy

At a government level, this is of equal importance, but in this case as a tool for driving innovation more effectively through a better understanding of the direct benefits behind the changes being promoted. The compliance-led approach that has become the accepted way to guide the industry towards 'best practice' solutions has resulted in a disconnect between those solutions and the reasons why they should be chosen. This is an opportunity to redress that balance, re-educate the industry, and get better outcomes without so much reliance on legislation and financial incentives.

Within all these different reasons for adopting and benefitting from such a decision-making tool lie the collaborations that need to be formed to help promote it. Mutually beneficial partnerships are the best way to spread the burden and spread the word, but ultimately there has to be ownership or at least endorsement, preferably at an industry level, to ensure the project's success. In an ideal world this would be the easiest part of the operation, but in a political environment where short termism trumps genuine policy making, this cannot be relied upon. Crowdfunding and opensource development are becoming increasingly popular drivers of innovations such as these that can benefit from both the finance and the publicity that such an approach offers. The break with conventional funding methods, predicated as they are on the capturing of IP, has been discussed already, where the need to reach a broad and open market can be stifled by the desire to capitalise too soon on a product's potential instead of capturing the market first and letting the financial benefits trickle in from the entire industry a little later. Small slice big pie wins over big slice small pie every time.

Conclusion – Three Tools in Context

The three tools discussed in this chapter have each been developed to fill a gap in how the construction industry currently operates. The first was to help the industry collect the evidence it needs to prove the benefit to be had in changing the way it operates. The second was to help the industry communicate with itself and better understand how each of its component parts might have a different perspective on where those benefits lie. And the third was to provide a vehicle for making better decisions based on the more complete and balanced information the first two tools would deliver.

How these tools can be made available is for the final conclusion, but before that, the next chapter brings all these approaches together to tackle the ultimate conundrum, our very special, possibly unique, and certainly self-inflicted housing crisis.

Bibliography

Climate Change Committee. (2022). *2022 Progress Report to Parliament*, June. www.the
    ccc.org.uk/publication/2022-progress-report-to-parliament/
Environmental Law Foundation. (2021). *Local urgency on the Climate Emergency?*
The Construction Innovation Hub. (2021). *Value Toolkit Overview Document* (Issue
    April).

8 Applying the Tools to the Housing Industry

In the world of Systems Thinking and Wicked Problems, the housing industry represents an extreme case within the broader construction industry. More than any other, it symbolises the conflicting demands of the technological, economic, societal and environmental factors that must be satisfied before any one of housing's many masters will be satisfied with the outcome. The government, the supply chain industries, the housebuilders, the design industry, the warranty industry and of course the end users are all impacted, if not party to the decisions that are taken throughout the long and tortuous process of delivering a home.

The evidence that 'not all parties are party' to the decisions being taken in this endeavour is starkly on show in the housing we are building. Even now, despite all the investment that is flooding in to finance the 'New Entrants', who are keen to re-imagine how we build our homes in the image of the manufacturing industry, little attention has been paid to what it is the homeowner or occupier actually wants to live in. In fact, little attention has even been paid to what the housebuilding industry wants from this move towards manufacturing, so confident are those promoting this latest intervention that the problem of how to build more for less can be easily resolved.

Solutions from a New Perspective

In this final chapter, some of the more recent panaceas put forward as solutions to our housing crisis are examined, and through applying the same holistic process of analysis from all perspectives – and in this instance there are plenty to consider – a conclusion is reached for how to resolve these seemingly intractable conflicts of interest that are preventing any substantive move being made away from the industry's status quo.

To follow the prescribed process, settling on the right question to ask would seem to be the preferred starting point for this endeavour, but in reality any questions relating to housing eventually lead back to the same conclusions, as all questions merely represent different entry points into the same constellation

DOI: 10.1201/9781003332930-11

of interrelated issues, causes and consequences. For now, the question chosen as that starting point is:

'Why have we got a housing crisis?'

Because we've had a housing crisis for some time now, there are many different opinions for why this is, all of which have some basis in perception or reality, but none of which can be said to be the sole reason, as some proponents would like to suggest. All of these need to be captured, however, before we

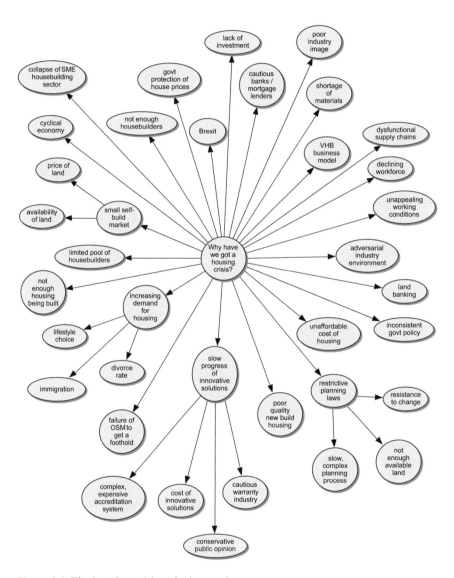

Figure 8.1 The housing crisis – the known issues.

can begin to test their credibility or measure their impact. This needs to be treated as an iterative process, as other causes will emerge through investigating those added at the outset and questioning why it is that they have arisen in the first place.

Clearly there is no shortage of explanations for why the industry is in the state it is, but we need to establish a hierarchy of causes and consequences and,

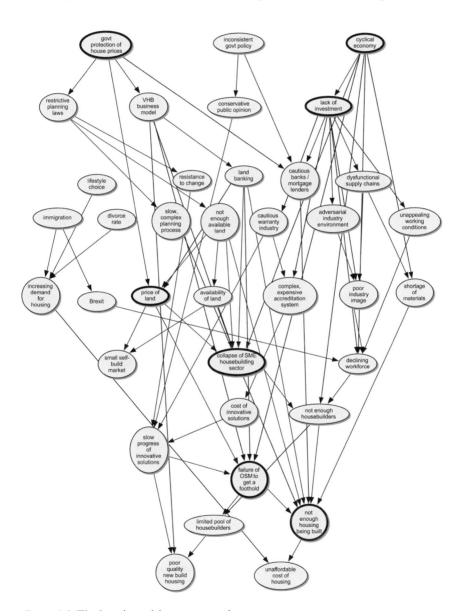

Figure 8.2 The housing crisis – causes and consequences.

within that, where to act if we are to tackle the root causes that are within our gift to influence.

To start at the bottom, the consequence of our housing crisis is not just the shortage of housing but the quality and unaffordability of what is being built. And this would appear to be due to the limited pool of suppliers, in turn due to the failure of the OSM industry to get a firm foothold in the marketplace, together with the more recent collapse of the mostly traditional SME housebuilding sector, which has been in steady decline since the 2008 crash (NHBC Foundation, 2017). There are multiple reasons for why both of these models have struggled to compete against the increasingly centralised volume housebuilding sector, but the recurring themes are access to land due to price and planning laws, and a cyclical economy that disincentivises any serious long-term investment in skills training or production plants. The volume housebuilder's business model, through the development of its 'survivalist shape', has become immune to all of these problems. It operates across economic cycles without such devastating losses, due to its policy of land banking from which a large proportion of its profits accrue, and its adherence to traditional on-site construction, requiring minimal investment in the assets of building. Furthermore, labour is seen as expendable, and 'let go' when the downturn begins. But when labour is in short supply at the beginnings of an economic upturn, their business model relies upon OSM companies to take up the slack, only to abandon them later on in that cycle when labour again becomes cheap and plentiful. In this way it has historically taken advantage of, and kept at bay, competing industries who, as newcomers invested heavily in off-site factories, must maintain a steady throughput of work, or, as smaller SMEs without the resources to invest in land, do not have the luxury of adding the uplift in land value over a ten-year period to their profit margins.

There were some sweeping generalisations in that analysis, set out here more as questions to be posed rather than as irrefutable facts. And there are indeed many other actors in this play, but it is the root causes that must be tackled first before turning our attention to the minor performances, most of which are in any case only there as a consequence of these more systemic issues. But as usual those root causes exist mainly within the political arena, and a decision must be made whether to challenge them head on or accept them as immovable barriers, and work within the constraints they define.

'No return to boom and bust' always felt like a bold statement, even coming from Gordon Brown (Lee, 2011), and with global events now impacting upon our economy with ever increasing frequency, it would seem reasonable to accept our cyclical economy as an uncontrollable given, and consider instead how we can best ameliorate its impact on our housing industry. A steady continuous housebuilding programme would certainly resolve a lot of the issues seen to be stemming from the stop–start dysfunctionality we currently have:

- More incentive to invest in skills and training
- Reliable, continuous work attracting more, better qualified labour into the sector

- Better quality construction from a regular qualified workforce
- A chance for OSM to become profitable over a longer period than one economic cycle
- A broadening of the supply base to include more SME businesses and a greater variety of business models
- More competition, more housing built, lower prices for house buyers and renters.

But how can this be achieved? The right government policies would need to be enacted, but only ones that would be seen as beneficial at a governmental level as well as at an industry and a public level. The pragmatic approach would be to assume the volume housebuilders will continue to follow their preferred model until it no longer makes economic sense for them to do so. That model is to provide the type of housing that will generate the most profit – family housing, as opposed to what is most needed – social housing. This would suggest that the provision of social housing should be seen as the responsibility of the state and should be prioritised during recessionary periods, utilising the capacity of OSM businesses who would otherwise fail. The volume housebuilders' traditionally trained labour force should also be utilised during recessionary periods, but for retrofitting programmes, where their skills are far more valuable and transferable than in the production of factory-built housing. In addition to this, those at the end of their careers and considering retirement should be encouraged into teaching positions to pass on these skills to the next generation entering into the complex and growing retrofit market. During boom periods, the OSM industry could continue to provide the social housing requirement for the volume housebuilders' developments, rather than competing against them, and the retrofitting programmes could either continue, or their workforce could transition across to the traditional housing sector at the peak of the boom, rather than them recruiting untrained labour and seeing the quality of their builds plummet, as has cyclically happened before (House of Commons, 2016).

With this smoothing out of housing supply, there would be an increase in investment, skills and productivity, and more homes would be built by more companies. OSM would become a more stable, trusted platform and the warranty companies would become less cautious as fewer, better designed solutions begin to emerge and be seen to cost less and perform better than their traditional counterparts. Only then would the volume housebuilders genuinely consider transitioning to MMC as opposed to using their products to prop up their business model when it suits them to do so. Their business model is the one to beat, and until what is being offered by the OSM industry is better than the status quo, the status quo will remain.

Creating the Environment for Change

The volume housebuilders are often portrayed as the villain of the piece in this story, but we have a free market economy, and we should not expect the businesses within that system to behave any differently than they do. It is not their job to

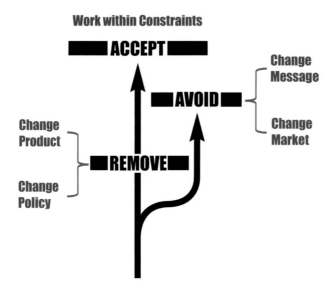

Figure 8.3 The options – confronting the barriers to bring about change.

help promote MMC for the greater good of the economy or to help reduce the industry's carbon emissions. Instead, we should be using them as the barometer against which we assess how well we are doing at replacing their product with something better. We've accepted that this is not a level playing field, and that the volume housebuilders have created an advantage for themselves that cannot be matched without some radical policy changes from above, but before we resort to that, what else could be done from within to improve the OSM offer?

Change the Message

To understand just how powerful marketing can be, it is worth revisiting an earlier incarnation of OSM when the brick and block industry decided in the 1980s that the timber frame industry was becoming a threat. They commissioned a report into some particularly poor timber frame homes in the South West of England, and in 1983, an ITV documentary made by *World in Action* alleged that timber frame construction could not produce houses that would last, citing rot in the frames of nine-year-old homes on a Cornwall estate. The industry collapsed and took nearly 20 years to recover, despite the claims being largely discredited soon after. But in Scotland timber frame continued to be used and grew to become a mainstream solution – because the programme was never aired there.

But messaging can also be too positive. The push for OSM has largely been driven by government-backed campaigns promoting the benefits of manufacturing housing off-site in terms of productivity boosts, environmental gains, health and safety, even cost savings, all of which are either debatable claims, or of no direct interest to those being asked to transition to this new way of building. As argued in Chapter 7, changing mindsets requires a more balanced approach that recognises the reality of the decisions being made and the need to understand the pros and the cons before jumping. Taking this more balanced approach not only provides a truer picture for the customer, but also exposes some of the flaws in the proposition that so easily get overlooked in the rush to promote a product regardless of its shortcomings. This is most pronounced in the volumetric sector, trumpeted by its proponents as the ultimate solution for manufactured housing. Some, however, not least the panelised timber frame industry, would disagree for the following reasons:

- The loss of flexibility in the design of the end product reduces the size of the market it can appeal to.
- The range of site conditions where volumetric housing can be used is limited, with sloping sites often becoming unviable.
- The density that can be achieved can also be affected, especially on smaller, more complex sites.
- The size of the homes being delivered also limits the sites that can be accessed and often introduces additional costs for police escorts, cable removal, cranage, ground strengthening, road closures, timetabling, delays, rescheduling costs, etc.
- The accuracy required in construction to allow for the seamless matching of ground and first-floor units often results in the need for on-site remedial works.
- Planning requirements interfere with the standardisation of the production line, requiring special brick slips (leading to delays in manufacturing), changes to roof pitches (additional delivery costs as roofs may no longer fit under bridges and have to be delivered and craned on separately), changes to window positions for overlooking issues (requiring changes to hoisting positions and external wall structures).
- Regional differences in regulations for mobility access, etc. can change room sizes and require non-standard design solutions.

The tendency to dismiss these concerns can lead to disillusionment during loss-leader demonstrator site projects that leave both the customer and the provider unwilling to repeat the exercise, or encourage anyone else to replicate the experience. But if these are genuine shortcomings, surely the messaging is failing to address them. Or perhaps the product itself needs to be revisited.

Change the Product

There are clearly benefits to manufacturing housing in factory conditions – more sustainable methods of construction, more control over accuracy and quality of build, less waste, better safer working conditions, proximity to work-force, equal gender opportunities, etc. The problems listed above, however, are almost entirely related to the loss of flexibility due to standardisation and size. But there is a relatively simple solution to this once those issues are recognised as barriers to adoption that need to be addressed.

Flexibility and standardisation might sound like opposites, but not once the purpose behind that standardisation and how it can be achieved is fully under-stood. Standardisation is being driven by the need to boost productivity in an industry that has been flatlining for over 20 years. Methods of construction need to change, but the status quo has become too powerful and embedded to dislodge, and in the housing sector, the imbalance caused by land banking too great a financial advantage to compete against. But specifically for housing, standardisation alone is not the solution. It has a role to play, and in other sectors, such as education, there is considerable scope for this to reduce the build cost of schools at a national level with little detrimental impact or the need for endless variations to encroach upon and dilute the benefits of such a process. Housing, in comparison, is far more of a social undertaking where the individual needs of clients, sites and occupiers all work against the goal of standardising the end product.

The recent demise of some of the very highly financed companies that moved into this space (Brown, 2022) would seem to suggest that the barrier of overly standardised solutions has been hit head on. What has not been under-stood, perhaps, is that the real benefits of standardisation lie in the process as much as they do in the product. The analogy of the automotive industry has by some been taken too literally, and with little attempt to discover its breaking point. Production lines developed for the manufacturing of a range of cars require investments in the tens of billions (Mijatovic, 2022), but they are for fully automatable processes and for the manufacturing of millions of cars that can be driven on any road in the world. The number of identical houses that might ever be wanted let alone be appropriate for enough sites to take them is of a different order of magnitude, to say the least. The system developed for the 'manufacturing' of housing needs to reflect that.

The aim needs to be towards developing a standardised 'kit of parts' that can then be used to create a variety of different homes, but following a standardised process that ensures that the method of construction is always appropriate without the need for additional design, material or structural input. This is the Platform Design Approach, and it needs to be developed for each sector of the construction industry to optimise building for all building types. Some sectors will find their needs coincide enough to justify sharing the same 'platform', whilst others, like housing, will probably need to develop their own strategy that reflects the far greater emphasis that will need to be placed on flexibility than will be the case in other sectors.

So to return to the volumetric housing model and how this could be improved upon, and using Platform Design's 'kit of parts' approach, at a macro level a house is effectively a kit of parts, with each part being a room.

Each of those rooms could be transported separately, removing the many problems associated with the size of a whole house product. But why should they be? Most of a house at the construction stage is made up of walls, floor and ceilings enclosing empty space. Far better to deliver those rooms 'flat-packed' and only build off-site the rooms with contents – kitchens, bathrooms, utility and services – as volumetric units. If those parts of the house can be combined and standardised into a range of options, and limited to a size that can be easily transported on a regular truck, the rest of the house can be designed flexibly around the site's and the end user's requirements. That is the basis for a housing solution that has now been fully developed and proven here as an outcome of Systems Thinking, and, given a more universal acceptance of such an holistic approach to problem solving, would no doubt have been developed and rolled out long ago.

Change the Market

There is another barrier to adoption, however, also experienced by the volumetric housing industry, and also resolved by this hybrid volumetric / flat-packed approach. Again, it focuses on the process, not the product, and it is of particular interest to the volumetric industry's key market, social housing. A better understanding of the volume housebuilders' business model would have helped the volumetric industry understand why their solution is of little interest to them, which, to avoid repetition, is all of the reasons given at the beginning of this section. But in addition to these is the problematic need for the finished homes to be delivered at a pace to suit the manufacturer rather than the client's unpredictable rate of sales. To maximise profits, volume housebuilders control the build-out rate to match demand, and they also build a variety of price points into their housing estates, neither of which dovetail with the volumetric business model.

The social housing market is very different. The housing is for rent, which means it is wanted as soon as possible to maximise rental income. The requirements are more standardised and around higher nationally prescribed standards, making social housing potentially cross-tenure and a better fit for a high-performance manufacturer with a limited range of house types. But more importantly, social housing providers tend to favour solutions that benefit the 'local pound' and create jobs within their district (White, 2020) – which is a major barrier for a company manufacturing and delivering their homes from elsewhere. The advantage of the hybrid approach is in the split between manufacturing in a factory setting – locally or regionally – and fabrication of the component parts on-site, a semi-skilled operation that can be carried out by local labour without needing to source professional tradesmen.

Flexibility in the manufacturing process also extends to the delivery of the product. Because the volumetric component is manufactured to the same system as the panelised components, where that volumetric unit is constructed can be decided by the scale of the project and the availability of factory space, skilled labour or even the need to set up a training centre. The units can be manufactured on-site, off-site or near-site in a flying factory. Increasing these options increases the size of the market that the product can appeal to, so that the all-important throughput on which all OSM operations depend can be maintained.

Change the Policy

Already the potential for success has been greatly improved by the suggestion of changing the messaging to reflect the practical realities of OSM, which would gain trust, but would also expose the need for an improvement in the product's design to reduce the barriers to adoption not being openly discussed. Furthermore, having addressed that, the most appropriate market for that improved product to be sold in to would become clearer. On the policy front then, it has been accepted that expecting to put an end to boom and bust might be a step too far, but the question of land value is still something that could and should be addressed if we are to create a level playing field for this and other innovative solutions.

Land value is in fact the elephant in the room, the barrier at the end of the road, the ultimate reason why the construction industry, and the housing industry in particular, is in a near existential crisis. Whilst we have a shortage and a chronic undersupply of housing, house prices will always be defined by the 'residual land value' calculation.

This calculation, central to the volume housebuilders' business model, says that the amount paid for developable land is the difference between the sale price of a property (including profit) and the build cost. There are two protected values in this calculation, the profit margin and, whilst there remains an undersupply of housing, the price at which property can be sold. So if MMC solutions manage to reduce the build cost of our homes, instead of the sale price reflecting this by coming down, the difference is absorbed by the land sale price going up. This has been the formula on which house prices have been predicated since the 1961 Land Compensation Act came into existence, an Act of Parliament by which land owners have benefitted through unearned windfall profits to the tune of over £10bn a year and rising. All other European countries capture a large percentage of this uplift in land value through various compulsory land purchase mechanisms, as we did prior to 1961, when we built 21 new towns on land purchased at a reasonable price that then accounted for around 2–3% of the total build cost (Bentley, 2017). The land cost now for an equivalent New Town development would account for over 60% of that cost, which is why we will not be building any, despite the policy being rolled out repeatedly by successive governments to placate the electorate.

The effect this has on innovation is to take the financial benefit in reducing build costs out of the loop, and with it any incentive to innovate. Even the short-term fix of reducing our house sizes – now the smallest in Europe and 80% smaller than the largest (Morgan & Cruickshank, 2014) – to maximise

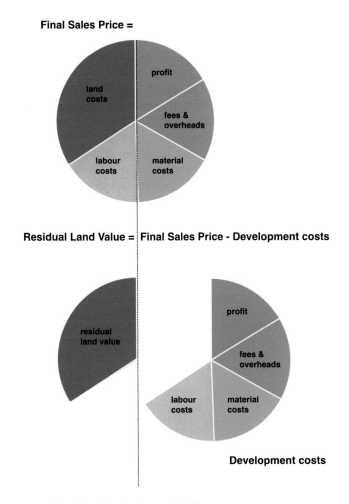

Final Sales Price =

Residual Land Value = Final Sales Price - Development costs

Development costs

Figure 8.4 Residual land value calculation.

profits on purchased land only serves to increase the price at which it is still profitable to buy that land going forwards. The corollary of this, however, as was proven by the Code for Sustainable Homes targets, is that when policies are put in place that drive a predictable increase in regulatory requirements and subsequent build costs, the price paid for land decreases to protect profits rather than increasing the sale price above the norm.

This is undoubtably a problem that needs fixing, but as has already been stated, it is a problem that has been known about and debated in Parliament for over 100 years. There is no shortage of policy papers and think tanks recommending ways in which government policy could be revised to implement changes and bring about a more equitable housing delivery system, and they all recognise the underlying issue of land value and how 50 years of unfettered speculation has resulted in our present predicament (Aubrey, 2015). It would seem disingenuous therefore to suggest that the government lacks a similar level of understanding, leaving only a lack of motivation or a lack of ability to explain their failure to act on this front. As a line of research, reaching conclusions at this level becomes increasingly difficult, as straightforward answers are hard to come by. Motivation and ability also become blurred as both are clouded by the often more powerful need to get re-elected. But whilst there is endless speculation as to what political purpose might be served by preventing the adoption of a more equitable system of land procurement, ultimately the status quo continues to prevail. The longstop, as already stated, would seem to be the point at which less than half the voting public own property, an increasingly predictable and fast approaching date, but there are some shorter-term policy measures that could be considered whilst we wait for that demographic endgame to play out. For the government in power, those short-term policies currently consist of unproductive measures designed to 'prop up' home ownership and property values – two conflicting objectives – by offering more and more incentives for first-time buyers, rather than attempting to deal with the root causes that are known to exist. So here, to end with, are two Systems Thinking derived suggestions for what could be done in the short term, based upon the need to satisfy the motivations of all parties, and for there to be any realistic chance of a shift in policy.

Hump Funding

Way back in Chapter 4, it was suggested that, alongside recognising the need to consider all stakeholder perspectives, it would also be pragmatic to accept that some sectors would have more influence than others over the decisions being made. In this instance there is a power play between government and the volume housebuilders, bordering on an unholy alliance. The volume housebuilders' position is relatively straightforward, although successive housing ministers seem to have struggled to grasp this (Shoffman, 2018). Their motivation is to maximise their profit and to satisfy their shareholders. Housebuilding for them is a process necessary primarily to realise the uplift in land value that turning agricultural land into developable land offers. Any change to that model that threatens their profit margins – land reform, competition from off-site manufacturers, regional planning restrictions, tighter building regulations – will therefore be resisted. Their ability to do this comes from their lobbying power, bolstered by the fact that less than a dozen individuals from that sector are responsible for over 20% of the Conservative party's funding (Williams, 2021).

    The government, on the other hand, has a far more complex portfolio of conditions to meet. It has committed to increasing the delivery of housing to meet a cumulative shortfall that is now seriously impacting on our economic performance (Gibb, 2023). It also needs to protect house prices from a collapse that would be equally damaging to the economy, but would also alienate the bedrock of its electorate. Walking this tightrope whilst not upsetting its major donor has led to some placatory deals that have not delivered. Decisions to row back on sustainability targets, reduce planning restrictions and incentivise first-time buyers in return for increased output from the volume housebuilders have done little more than boost their profit margins whilst exacerbating many of the problems that still need to be resolved, not least the concentration of our supply base into fewer and fewer national providers. For the public to see how much better their housing could be, they need to see built examples of this at scale, but the MMC solutions the government needs to progress – not least to satisfy its other statutory requirement of achieving a net zero economy – will not be built without there being other players in the market. But if that has been ruled out as a fix that the government cannot bring about without first destroying the only viable business model in town, the solution has to lie in getting the volume housebuilders to build to that higher standard themselves.

The route to this is in using the residual land value calculation in its positive incarnation, but at the same time, temporarily compensating the volume housebuilders for their lost profits. If asked, the volume housebuilders will say they don't have a problem with building their homes to the upcoming Future Homes Standard as long as they can maintain their profit margins. But it inevitably will cost more to incorporate the changes being proposed, and they have already bought land at a price that can only support current build costs. If the specification of future builds is known, however, and more importantly, can be relied upon not to be reversed, future land will be purchased for less to compensate for this, meaning that all the government needs to do is provide some relatively affordable 'hump funding' to compensate the volume housebuilders for the additional build costs of housing to be built on land already purchased.

Whilst that may seem unpalatable to some, it will ensure the continuation of housing delivery but to the new standards required. And since the cost of retrofitting new homes built to current standards now has been calculated at five times the additional cost of building them to the new standards at the outset (Currie & Brown & AECOM, 2019), this should make sound economic sense. Ideally this would be implemented alongside a conditional planning policy that would prevent the practice of long-term land banking in the future so that the playing field could be levelled for SME builders.

The Pre-App Gatekeeper

That leads on to the second proposition, which is for a radical but mutually beneficial policy change to the planning system. The residual land value calculation is something that exists mainly within the housing sector. In other sectors, the resistance from developers to the introduction of sustainability and net zero measures will need to be tackled through other roots. Our planning system, much like our tax system, has been tweaked and added to over many years and has become unwieldly, but has proven very difficult to strip back or reinvent to the degree needed to make it function as originally intended. The addition of sustainability targets to be policed by an already stretched and underfunded planning system has resulted in some developers being able to pay lip service to these new regulations and force through developments that, at best, only meet current standards and certainly not future aspirational targets – this despite over half the country's authorities declaring climate emergencies and committing to hit net zero by as early as 2028 (Notttingham City Council, 2020), more than 20 years before the national statutory requirement.

Achieving anything close to carbon neutrality in any building requires considerable effort and expertise, and whilst developers may often pay for the research needed to work out how best to minimise their emissions, they invariably baulk at the costs involved in implementing the research's findings. Instead, the preferred route is to pay for a report that purports to hit the required targets and let the planning authority have to prove that it does not, and do this within the statutory time period allowed for a decision. Both the time allowed and fee payable for planning applications are fixed nationally, meaning that, once submitted, the clock is running, allowing little room for manoeuvre when confronted with a large complex scheme. Even with more time, it is still unrealistic to expect planners to have the in-depth knowledge needed to assess whether or not a proposed solution is genuine or just greenwash. Neither is it realistic to expect them to be able to buy in that expertise on their restricted budgets to assess every scheme submitted.

The proposed solution therefore is to deal with this at the pre-application stage. This is currently a service offered by most planning authorities for a nominal fee to assess the likelihood of a scheme being acceptable for a full submission, but it is the only stage in the process where local authorities have any real control, both over the price and what can be offered in return. The cost of the planning process is in reality of little significance to developers, whereas the delays that can result from it when a scheme is contested can be financially crippling. Even the cost of complying fully with sustainability targets, once compelled to do so alongside all other developers, is not the financial Armageddon it is sometimes portrayed to be. It is the additional cost over that of a competing scheme that manages to dodge those costs that is problematic.

The proposal is therefore to make the pre-app stage a condition of making a full application, and to charge enough for that pre-application to cover the costs of a qualified third-party professional body that can assess whether or not the carbon emission claims being made are genuine or not. Only after a successful assessment can that scheme pass the gatekeeper and enter into the full planning process, with the guarantee of a straightforward passage through, on sustainability grounds at least, within the time allocated.

Both of these proposals involve additional upfront costs, either for government or for the developer, but in both cases, those costs are considerably less than the alternative remedial costs that would otherwise begin to accumulate. This is not an unusual scenario, but the important difference is that the decision to act now rather than pay later would be put back into the hands of the enforcing government body who can and should be taking that longer-term perspective. If governments cannot manage that responsibility, there is little hope of any business in a free market economy managing it for them.

Bibliography

Aubrey, T. (2015). *The challenge of accelerating UK housebuilding: A predistribution approach* (Vol. 44). www.policy-network.net

Bentley, D. (2017). *The land question*. Civitas.

Brown, C. (2022). Growing pains or cause for concern? What recent financial failures mean for the modular market. *Housing Today*, June. www.housingtoday.co.uk/in-focus/growing-pains-or-cause-for-concern-what-recent-financial-failures-mean-for-the-modular-market/5117864.article

Currie & Brown, & AECOM. (2019). The costs and benefits of tighter standards for new buildings. *A Report for the Committee on Climate Change*, February. www.the ccc.org.uk/wp-content/uploads/2019/07/The-costs-and-benefits-of-tighter-standa rds-for-new-buildings-Currie-Brown-and-AECOM.pdf

Gibb, K. (2023). Doing UK housing policy well is hard, but it has never been more important. *UK in a Changing Europe*. https://ukandeu.ac.uk/doing-uk-housing-pol icy-well-is-hard-but-it-has-never-been-more-important.

House of Commons. (2016). *More homes, fewer complaints*. July.

Lee, S. (2011). Boom and bust: The politics and legacy of Gordon Brown. *Local Economy: The Journal of the Local Economy Policy Unit*, *26*(6–7), 619–622. https://doi.org/10.1177/0269094211417357

Mijatovic, S. (2022). A detailed look at the Volkswagen Wolfsburg manufacturing plant. *HotCars*, March. www.hotcars.com/volkswagen-wolfsburg-plant-detailed-look/

Morgan, M., & Cruickshank, H. (2014). Quantifying the extent of space shortages: English dwellings. *Building Research and Information*, *42*(6), 710–724. https://doi.org/10.1080/09613218.2014.922271

NHBC Foundation. (2017). *Small house builders and developers: Current challenges to growth*, April. www.nhbcfoundation.org/wp-content/uploads/2017/04/NF76_WEB.pdf

Nottingham City Council. (2020). *Carbon Neutral Nottingham: 2020–2028 Action Plan*. www.nottinghamcity.gov.uk/media/2619917/2028-carbon-neutral-action-plan-v2-160620.pdf

Shoffman, M. (2018). Sajid Javid sets out his housing dream but has Help to Buy become a nightmare? *Property Industry Eye*, January. https://propertyindustryeye.com/forget-the-housing-dream-has-help-to-buy-become-a-nightmare/

White, R. (2020). Briefing: Social housing & Britain's housebuilding recovery. *Shelter*, June. https://england.shelter.org.uk/professional_resources/policy_and_research/pol icy_library/briefing_social_housing_and_englands_housebuilding_recovery

Williams, M. (2021). 20% of Tory donations come from property tycoons. *Open Democracy*, July. www.opendemocracy.net/en/dark-money-investigations/20-tory-donations-come-property-tycoons/

Conclusion

To quote Albert Einstein for a second time,

> Insanity is doing the same thing over and over and expecting different results.

This is possibly an unfair statement to level at the many and varied attempts that have been made over the years to understand and fix the construction industry's problems, because each government or industry-funded report has focused on a different area of concern and to different degrees of success. It is also unfair because the remit of these reports does not extend beyond making recommendations into how those recommendations should be enacted. The insanity, if it exists, is in continually throwing these balls and expecting someone to catch them.

Somewhere the ball is being dropped, but rather than extending that analogy any further, we need to work out where the problems lie in the real world and fix them. The suggestion that has been made here – and it is only a suggestion – is that the problem is predominantly one of messaging, and a failure to translate these recommendations into a language that those who need to take action will understand. The problem may well run deeper and be due to a lack of understanding of the problems being confronted from the perspectives of those confronting them, but these are classic 'cause and consequence' relationships. The bottom line is that if we don't fully understand the problems, it's unlikely we're going to understand how to fix them.

> If I had an hour to solve a problem I'd spend 55 minutes thinking about the problem and 5 minutes thinking about solutions.

Yes, it's Albert again, eloquently stating the obvious. Spending more time understanding a problem means you're far more likely to arrive at the right solution. Even though Einstein's ownership of that truism has never been proven, it's certainly worthy of him, and is as relevant to constructional problems as it

DOI: 10.1201/9781003332930-12

is to problem solving in general, and has to form the foundation for what the construction industry – and government – needs to do differently.

All the propositions put forward in the second section of this book have been ones that were born out of taking this approach to known problems that exist within the construction industry, and many of the conclusions reached were arrived at by merely applying an existing knowledge of how other sectors operate, but interrogating that knowledge through a systematic process of analysis from multiple perspectives. The conclusions themselves still require validation from those sectors, but the point is, considering those alternative perspectives is not such an onerous task, and certainly not a thankless one once the benefit in doing so has been appreciated. The payback will be in the solutions that emerge. They will stand a far greater chance of success, they will last far longer and will ultimately be of far more benefit to many more people. Or, in the language of Systems Thinking, if we don't all win, nobody wins.

Index

adoption of policies 32
adversarial industry 56
advisory bodies 66
affordable housing obligation 38
architect as master builder 21
Asian market 72
authenticity 50
automation 64
automotive industry 111, 194
awareness 50

barriers to change 48, 97, 190
behavioural change 174, 176
binary choices 98
biodiversity loss 20
Black Market 110
BLP Insurance 39
bonfire of red tape 14
boom and bust 55, 60
Bossom, Alfred 1
box ticking compliance 117, 148
BREEAM 121
bricklayers 42
British Research Establishment 121, 161
Brown, Gordon 190
Build Back Better 3
build out rate 36, 41, 195
Building Information Modelling 23, 56, 71, 128
building trades 19
Buildoffsite 138

carbon neutrality 30
carbon offset funds 103
carrots and sticks 36
CAST consultants 32
causes and consequences 30, 62, 115, 169
change management 76, 85, 148
change the message 174, 184–5, 192

change the product 174, 193
change washing 76
changing mindsets 46
Cherry picking 33
Churchill, William 115
Climate Chane Committee 123
climate change 30
Climate Change Act 2008 119
climate emergency declarations 141, 176, 200
CLT 14, 98
Code for Sustainable Homes 123, 197
coercion 41
Collaborating for Change 129
collaborative working 22, 44, 70
common denominators 51
Community Infrastructure Levy 37
comparative assessment 50
competitive tendering 67
complacency 46
complexity 55, 73, 77
confidence levels 155
conflicting motivations 172
Construction Industry Training Board 68
Construction Innovation Hub 138
Construction Leadership Council 33, 39
Construction Sector Deal 33, 43
convenience 146
cost benefit 24
cost centres 149
cradle to gate 130
cradle to grave 130
critical mass 71
Crossing the Chasm 40
custom self-build 93
cyclical economy 44, 55, 60, 71

data gathering 149
decision making tree 88

defender markets 100
delivery base 155
democratic realities 62
demonstrator sites 152, 193
digital twin 153
disrupter business models 17
disruption to working practices 156
disunity 55, 77
dominant sector 90
Donations, Tory party 115
dynamic modelling 120

early adopters 40
ecology 54
economic resilience 20
economies of scale 144
Egan, Sir John 111
Einstein, Albert 31, 202
Elective Autocracy 14
Embedded knowledge 178
embodied carbon 20
Environmental Product Declarations 130
EPC ratings for offices 4
equity 54
ethical priorities 149
Exclusivity 92
exogenous barriers 57

Farmer Review 33
feasibility 50, 97
feedback loops 80
fire risk 101
Fixing Our Broken Housing Market 32, 35, 122
flexibility of solutions 156, 194
flying factory 196
fragmentation 65
fuel poverty 174
Future Homes Standard 199

government agendas 13
Government procurement 39, 66
Grand Challenges 44
greenwashing 200
Grenfell 14, 101, 118
gut instinct 52

health and safety requirements 20
heat pumps 176
Help to Buy 117
hierarchy of barriers 45
hierarchy of influence 14
Hockerton Housing Project 92
housing crisis 188

hump funding 198
hybrid volumetric/flatpacked housing 195
hydrogen 176

Impartiality 62
Implementation 33, 111, 147, 175
Industrial Strategy (Construction 2025) 33, 39
Industrial Strategy Challenge Fund 43
industry bodies 161
industry drivers 11
inelastic attention span 52
infrared heating 174, 176
intellectual property rights 125, 185
intervention strategies 112
interventionism 60
investment in R&D 110

job security 73

Kerkstoel 94
key motivating factors 90, 102
kit of parts 133, 194
knowledge, motivation and ability 45, 48

labour supply 190, 194
Laing O'Rourke 66
land banking 17, 42, 190, 199
Land Compensation Act 196
land reform 115, 196
large format blocks 55, 77
Latham, Michael 111
length, breadth and depth 58
Letwin, Oliver 32, 41–2
LEXiCON 161
Lobbying government 15
local pound 195
Localism Bill 62
London Borough of Hackney 14, 98

macro-economy 57, 69
market absorption rate 42
market subdivisions 26
messaging 106, 148
metrics 49, 137, 147
migrant labour 70
migrant workers 110
misinformation 50
missing perspectives 11
Modernise or Die 1, 33
Morrell, Paul 129
motivating factors 10
mutual benefits 91, 185

New Entrants 66, 131, 187
New Towns 62
non-negotiables 90
NYMBYism 37

offsetting 144
106 agreement 37
opensource development 126, 184
opportunism 67
oversight 21, 133

Passivhaus 120
path of least resistance 86
payback period 54, 142
perception 52
performance gap 153
piling 97
planning application fee 123
Planning Authorities 122, 199
platform design approach 125, 133, 194
poor industry image 73
positive disruption 33, 100
pragmatic idealism 54, 118, 159
price points 195
productivity 30, 64, 125
productivity levels 43
proportionality 163
public perception 101

qualitative assessment 151
quantity surveyor 21
question based problem solving 86
Question based problem solving 14, 165
quick response heating 176

radical incrementalism 85, 113
remove, avoid, accept 59
rental housing 63
research and development 72
residential building 24
residual land value calculation 117, 196
retrofit 140, 166
risk 55, 67, 94, 156
root causes 35, 62, 190
Route to Compliance 161

Scimago 115
scope 58
security of supply 101
self build housing 17
services engineer 21
single point perspective 42
SMEs 110

Social Housing Decarbonisation Fund 142
social housing provision 17, 175
speculative house builders 63, 175, 191
stakeholders 9, 35, 46
Standard Assessment Procedure 91, 120
standardisation 194
static modelling 120
status quo 57, 152, 191
structural engineer 21
subjectivity 50
supply chain companies 18, 125, 128
survivalist shape 35, 58, 124, 190
symptoms 59
systems engineering 81
systems thinking 76

target driven policy 111
tax incentives 174
tenure diversity 63
thermal mass 94
three pillars of sustainability 53
Timber First Policy 14, 98
timber frame construction 192
time poverty 119
top down delivery 110
trade associations 18
transitioning 119, 149
trojan mice 85

uncertainty 57, 65, 73, 77
uncontrolled variables 153
undersupply of housing 24
unintended bias 46
unintended consequences 176

value engineering 21
Value Toolkit 137, 147, 161
vernacular architecture 109
vertical integration 66, 131
viability 50, 96
volatility of the market 57
volume housebuilders 17
volumetric housing 111, 193

warranty industry 101
wealth tax 117
weighting 95, 157
White Papers 31
Wicked Problems 14, 84, 97
windfall profits 117
window of acceptable discourse 59, 61
workforce size 69
World in Action 192